Preventing Workplace Incidents in Construction

The construction industry is vital to any national economy; it is also one of the industries most susceptible to workplace incidents. The unacceptably high rates of incidents in construction have huge socio-economic consequences for the victims, their families and friends, co-workers, employers and society at large. Construction safety researchers have introduced numerous strategies, models and tools through scientific inquiries involving primary data collection and analyses. While these efforts are commendable, there is a huge potential to create new knowledge and predictive models to improve construction safety by utilising already existing data about workplace incidents. In this new book, Imriyas Kamardeen argues that more sophisticated approaches need to be deployed to enable improved analyses of incident data sets and the extraction of more valuable insights, patterns and knowledge to prevent work injuries and illnesses.

The book aims to apply data mining and analytic techniques to past workplace incident data to discover patterns that facilitate the development of innovative models and strategies, thereby improving work health, safety and well-being in construction, and curtailing the high rate of incidents. It is essential reading for researchers and professionals in construction, health and safety and anyone interested in data analytics.

Imriyas Kamardeen is Professor of Construction Management at Deakin University, Australia. He is also the Editor of the international journal *Construction Economics and Building*. His research and teaching interests predominantly lie in Construction Informatics and Workplace Health and Safety. He has authored two books previously with Routledge: *OHS Electronic Management Systems for Construction* (2013), and *Fall Prevention Through Design in Construction: The Benefits of Mobile Computing* (2015).

Spon Research

Publishes a stream of advanced books for built environment researchers and professionals from one of the world's leading publishers. The ISSN for the Spon Research programme is ISSN 1940-7653 and the ISSN for the Spon Research E-book programme is ISSN 1940-8005

Preventing Workplace Incidents in Construction

Data Mining and Analytics Applications

Imriyas Kamardeen

Routledge
Taylor & Francis Group

LONDON AND NEW YORK

First published 2020 by Routledge

2 Park Square, Milton Park, Abingdon, Oxon OX14 4RN

605 Third Avenue, New York, NY 10017

Routledge is an imprint of the Taylor & Francis Group, an informa business

First issued in paperback 2021

Publisher's Note

The publisher has gone to great lengths to ensure the quality of this reprint but points out that some imperfections in the original copies may be apparent.

British Library Cataloguing-in-Publication Data
A catalogue record for this book is available from the British Library

Library of Congress Cataloging-in-Publication Data
Names: Kamardeen, Imriyas, author.
Title: Preventing workplace accidents in construction : data mining and analytics strategies / Imriyas Kamardeen.
Description: Abingdon, Oxon ; New York, NY : Routledge, 2019. | Series: Spon research | Includes bibliographical references and index.
Identifiers: LCCN 2019012684 | ISBN 9781138087453 (hardback) | ISBN 9781315110462 (ebook)
Subjects: LCSH: Building–Safety measures–Data processing. | Building–Accidents–Psychological aspects. | Psychic trauma–Prevention. | Data mining. | Quantitative research.
Classification: LCC TH443 .K363 2019 | DDC 690.068/3–dc23
LC record available at https://lccn.loc.gov/2019012684

ISBN: 978-1-138-08745-3 (hbk)
ISBN: 978-1-03-217791-5 (pbk)
DOI: 10.1201/9781315110462

Typeset in Goudy
by Wearset Ltd, Boldon, Tyne and Wear

Contents

Disclaimer

This book uses data supplied by Safe Work Australia in March 2016, which were compiled from Australian workers' compensation organisations. The views expressed are the responsibility of the author and are not necessarily the views of Safe Work Australia or any of the Australian workers' compensation organisations.

Whilst the author has taken reasonable care in the preparation of this book, the use of the book is at your own risk and on an 'as is' basis. The author makes no expressed or implied warranty of any kind and will not be liable to you for any errors or omissions.

The author excludes any liability to you or anyone claiming through you for any loss, liability or damage whether directly or indirectly suffered or incurred by you or a third party arising from or in connection with the use of the book, including any information derived from it.

You should carefully review any recommendations or information contained in this book and obtain independent verification from a suitably qualified professional before acting on any of those recommendations. No responsibility is assumed by the author for any injury and/or damage to person or property arising from any methods, instructions or recommendations contained in this book.

Figures

Tables

Preface

Whilst the construction industry is an important sector in the national economy of any country, it has one of the worst records for workplace safety, health and well-being globally. For instance, the global construction industry employs only 7 per cent of the global workforce but accounts for 30–40 per cent of work fatalities, and one worker is killed every five minutes on a construction site. The dismal safety performance of the industry causes distressing socio-economic burdens for affected operatives, their families, construction organisations and society, globally. Improving safety, health and well-being in construction is therefore an urgent necessity to save lives and improve the socio-economic well-being of nations.

This book advocates a novel approach of applying data mining and analytics methods to improve health, safety and well-being in construction. The chapters in the book demonstrate systematically how different data mining and analytics methods may be applied on previous incident data to discover new knowledge and insights for incident prevention in the construction industry. Safety and well-being in construction is an international theme that concerns governments and WHS (Work Health and Safety) authorities around the world. This book will be relevant, value adding and will moot re-thinking of WHS management in construction internationally.

The book will equally appeal to both industry and researcher/academic communities. It can benefit builders, WHS practitioners in the construction industry, researchers, academics and learners in the following ways.

- It demonstrates how innovative construction companies can exploit contemporary techniques and theories to improve WHS performance and thereby business sustainability.
- It provides clear and easy-to-follow explanations for builders, WHS professionals, researchers and tertiary students on the application of data analytics and knowledge discovery.
- It functions as a valuable reference for safety analysts and researchers.
- It adds new knowledge related to construction WHS, leveraged by data mining and analytics methods.
- As far as the author is aware, no notable textbook on the application of data mining and analytics in construction is available. This book will address that gap and be a valuable reference source for construction academics and students.

Acknowledgements

I am indebted to various individuals and organisations for their valuable support, which enabled me to make this book a success.

First and foremost, I sincerely thank the almighty God for giving me the ability to write this book.

I am grateful to Safe Work Australia for providing me with the data required for the research that underpins the chapters of the book. The book would not have been possible without their support.

Special thanks go to Göran Runeson for carefully reading and providing feedback on the chapters of the book, which helped me improve their academic quality and rigour.

I would like to extend my appreciation to my wife, Zakia Rizvi, for her patience and continued support of my career, especially during the development of this book, and to my children, Mahdiyya and Imaad, for their love that fills my tired days with joy.

Finally, I would like to thank my extended family, friends and colleagues for their encouragements and support during my academic career.

Abbreviations

ABS	Australian Bureau of Statistics
ACT	Australian Capital Territory
ANZSCO	Australian and New Zealand Standard Classification of Occupations
BIM	Building Information Modelling
BLS	Bureau of Labor Statistics
CA	Correspondence Analysis
CHAID	Chi-squared Automatic Interaction Detection
CVDs	Cardiovascular diseases
DfS	Design-for-Safety
GDP	Gross Domestic Product
HSE	Health and Safety Executive
ILO	International Labour Organization
MCA	Multiple Correspondence Analysis
MSDs	Musculoskeletal Disorders
NN	Neural network
NSW	New South Wales
OHS	Occupational Health and Safety
OSHA	Occupational Safety and Health Administration
PPE	Personal Protective Equipment
PtD	Prevention-through-Design
RFIDs	Radio-Frequency Identification Devices
SPIs	Secondary Psychological Injuries
TOOCS	Type of Occurrences Classification System
UK	United Kingdom
US	United States
VR	Virtual Reality
WBV	Whole body vibration
WC	Workers' compensation
WHS	Work Health and Safety

1 Introduction

Dark side of construction

The construction industry is a vital sector in the national economy of any country; it creates the physical infrastructure essential for the functioning of the nation, provides jobs for a large number of people and contributes significantly to GDP. At the same time, it is one of the most vulnerable sectors for workplace incidents globally (Cigularov *et al.* 2010). In the Australian context, for example, Safe Work Australia (2015a) reported that 417 construction workers were killed due to work injuries over the 12-year period from 2002–2003 to 2013–2014 (2.24 fatalities per 100,000 workers). In 2013–2014, the construction industry accounted for 12 per cent of all work-related fatalities when it employed only 9 per cent of the national workforce. Moreover, the construction industry accepted an average of 12,600 workers' compensation claims per year for injuries and diseases involving one or more weeks off work over the five years from 2008–2009 to 2012–2013, equating to an average of 35 serious claims each day. Likewise, the UK construction industry recorded 35 fatalities in 2014–2015 (1.62 per 100,000 workers), the second highest among the main industry sectors (HSE 2015). The number of fatalities in the US construction industry in 2014 was 874 (9.5 per 100,000 workers), which is the highest among all industry sectors (BLS 2015). Construction fatality statistics in developing countries are much worse than these. For instance, Kheni *et al.* (2010) claimed that Ghana, a Sub-Saharan African country, recorded a fatality rate of 77.6 per 100,000 workers. In Turkey, official statistics for 2011 revealed that the construction sector accounted for 6.3 per cent of the labour force but 33.5 per cent (570 of 1700) of total fatalities for all industries (Gürcanli and Müngen 2013). Yoon *et al.* (2013) summarised that the Korean construction industry accounts for a fatality rate range of 18.9–39.7 per 100,000 workers and for the Taiwanese construction industry it was 13 fatalities (Cheng *et al.* 2012). The undesirable situation around the globe was summarised by the International Labour Organization that the construction industry employs 7 per cent of the global workforce but accounts for an excessively disproportionate 30–40 per cent of work fatalities; 2,100,000 workers are killed on construction sites every year (one worker every five minutes) (Murie 2007).

Table 1.1 Socio-economic consequences of construction incidents

	Tangible costs	Intangible cost
Victims	• Loss of salary • Reduction of professional capacity • Medical costs • Loss of time (due to medical treatments)	• Pain and suffering • Moral and psychological suffering (especially in the case of a permanent disability) • Lowered self-esteem, self confidence • Strain on relationships • Lifestyle changes
Family and friends	• Financial loss • Extra costs	• Moral and psychological suffering • Medical and family burden • Strain on relationships
Co-workers	• Loss of time • Increase of workload • Training of temporary/substitute workers	• Psychological and physical distress • Worry or panic (in case of serious or frequent incidents/cases of ill-health)
Employers	• Internal audit costs • Production loss • Damages to equipment and materials • Quality losses • Training of new staff • Technical disturbances • Organisational difficulties • Increase in production cost • Increase in insurance premium • Early retirement • Administration costs • Legal costs	• Presenteeism • Company image • Working relations and social climate
Society	• Loss of production • Increase of social security/welfare costs • Medical treatment and rehabilitation costs • Early retirement of workforce • Decrease of the standard of living	• Reduction of the human labour potential • Reduction of the quality of life

Source: adapted from European Union (2011, p. 8).

The unacceptably high rates of incidents in construction have huge socio-economic consequences for the victims, their families and friends, co-workers, employers and society in general. Table 1.1 illustrates the costs borne by these different groups. Several studies have been conducted to quantify these socio-economic costs of construction incidents in monetary terms. For instance, Safe Work Australia (2015b) estimated that the total cost of construction incidents in Australia during 2012–2013 was AU$5.84 billion, which is approximately 0.4 per cent of the GDP of the nation for that period, with an average unit cost of AU$117,180 per incident. It further reported generally for all industries for the period, that in terms of distribution, 77 per cent of the total cost was borne by workers, 18 per cent by the community and 5 per cent by employers. Likewise, in the UK, the total cost of construction injury and illness in 2013–2014 was estimated to be £0.9 billion. Waehrer *et al.* (2007) reported that construction incidents in the US in 2002 cost US$11.5 billion; the average cost of a construction fatality was US$4 million whilst a non-fatal injury was US$42,000. These statistics prove that the construction industry creates a distressing socio-economic burden globally. It is evident from the disturbing account above that preventing workplace incidents in construction is an urgent need.

Learning from past incidents

Improving safety and preventing incidents in the construction industry has been a top priority for the International Labour Organization and many Work Health and Safety (WHS) authorities around the world. Aligning with the priority, construction safety researchers have so far introduced numerous strategies, models and tools through scientific inquiries involving primary data collection and analyses. While these efforts are commendable, there is a huge potential to create new knowledge and models to improve construction safety by utilising already existing data about workplace incidents. Chua and Goh (2004) advocated that in order for the construction industry to improve its safety performance, it should learn from its mistakes and put the lessons learnt to good use. WHS authorities such as the Health and Safety Executive (HSE) in the UK, Occupational Safety and Health Administration (OSHA) in the US, and Safe Work Australia spent a significant amount of resources to collect data related to construction incidents. An enormous number of incident records exists with such authorities, in anecdotal form, with ample predictive and analytics potentials, which can be leveraged to develop incident prevention strategies (Panthi and Ahmed 2015).

The WHS authorities regularly analyse the data they collect and generate summaries, trends, charts, industry-specific statistics and comparisons to promote policy development initiatives. For example, Safe Work Australia publishes the following reports regularly, drawing from the incident and workers' compensation claims data it gathers:

- Lost time injury frequency rates for different industry sectors.

- Industry-based statistics that show an overview of the main causes of injuries and fatalities.
- Work-related fatality reports that summarise work-related deaths occurring in Australia.
- Work-related disease reports that identify concerning work-related diseases and their originating sources.
- Work-related injury reports that categorically summarise compensable injuries according to the nature of injury, incident mechanism, agency of injury and characteristics of workers.
- Costing of work-related injury and illness that provides an update of the 'human cost' of work-related injury and illness to the Australian economy.

Similar analyses are performed by HSE and OSHA. Construction researchers and commissioned research centres also have utilised these data or samples thereof for developing theoretical models. Notably, Dumrak *et al.* (2013), as a result of analysing workers' compensation data obtained from Safe Work South Australia, suggested a conceptual model to explain why some injuries are more severe than others. Haslem *et al.* (2005) studied 100 incident reports obtained from HSE and developed a model showcasing a hierarchy of causal influences in construction incidents. Huang and Hinze (2003), using data obtained from OSHA, revealed the characteristics and causes of fall incidents in construction. Chua and Goh (2010), utilising 140 construction incident cases obtained from the Occupational Safety Department of Singapore, developed a case-based reasoning system to aid construction hazard identification and safety planning.

Nonetheless, most of the analyses performed are either descriptive, which generate historical summaries, percentages and indexes (Ural and Demirkol 2008), or linear models that evaluate associations between incidents and possible causes (Karra 2005). The descriptive summaries produced by the authorities are helpful to some extent, but the data can be utilised to extract more insights. Likewise, the linear models generated by researchers do not represent the true nature of complex relationships between incidents and causes, and thereby restrict the possibility for explaining and interpreting workplace incidents in terms of the entire range of variables and effectively evaluating hypotheses aimed at curtailing those (Rivas *et al.* 2011). Therefore, more sophisticated approaches need to be deployed to enable improved analyses of these incident data and the extraction of more insights, patterns and knowledge to prevent workplace incidents.

Data mining and analytics for incident prevention

Data mining and analytics focuses on discovering new and interesting patterns, insights and knowledge in large datasets, differently from traditional statistical methods, which can be utilised for finding yet unrecognised and unsuspected facts (Äyrämö *et al.* 2009). Because of their powerful predictive capabilities, data mining and analytics approaches are used quite extensively in fields such as medicine, engineering, finance and business (Rivas *et al.* 2011). In recent years, the automobile sector utilised the approaches to mine traffic incident data to improve road safety (Beshah and Hill 2010; Äyrämö *et al.* 2009). It has been proven that data mining and analytics techniques can offer advanced predictive and explanatory capabilities for workplace risk and safety management (Martin *et al.* 2014). The mining and extractive industries have already started benefiting from them (Jacinto and Soares 2008; Silva and Jacinto 2012). However, it is an underexplored area for analysing workplace incidents in the construction industry (Hsueh *et al.* 2013). There are few studies that demonstrate descriptive analytics only; for instance, Cheng *et al.* (2012).

Researchers in other industries have indicated many benefits of applying data mining and analytics methods for investigating workplace incident data, namely:

- Data mining and analytics techniques such as decision rules, classification trees and Bayesian networks are reliable tools compared with classical statistical techniques in predicting and identifying factors underlying workplace incidents (Rivas *et al.* 2011).
- Data mining and analytics allows multivariate analysis of nominal variables with three, four or more categories, which is a challenge for statistical techniques, and provides a much more detailed and complete characterisation of incident patterns (Silva and Jacinto 2012).
- Data mining and analytics techniques enable studying the complex structure of interactions between all variables associated with incidents (Marques *et al.* 2014).
- Machine learning techniques, a family of data mining and analytics, have the capacity to investigate and automatically detect useful aspects of workplace incidents, which are hidden within the massive amount of information (Ciarapica and Giacchetta 2009). It is not necessary for the researcher to know the solid underlying relationships between input and output variables for model building; any relationship whether linear or nonlinear can be leant and approximated accurately through machine learning (Thipparat 2012).
- Incident related data are multidimensional, heterogeneous and may contain incomplete and erroneous values, which makes their exploration very challenging; yet, data mining and analytics methods are able to produce understandable patterns and useful results (Äyrämö *et al.* 2009).

Nevertheless, this area is almost a missed opportunity in construction safety research so far.

Data mining and analytics

The preceding section discussed the potential benefits of applying data mining and analytics methods on past construction incident data. A good understanding of the various methods of data mining and analytics is essential for their right applications to solve WHS problems. This section therefore provides brief outlines of the different data mining and analytics methods. Readers, however, should note that this chapter does not intend to provide detailed explanations of the field of data mining and analytics, which is beyond the scope of this book. Desiring readers may wish to refer to dedicated books on the subject. Among the example references for data mining and analytics are: Han *et al.* (2012), Tan *et al.* (2006), Kudyba (2014), Delen (2015) and Pierson (2017).

The concept of data mining and analytics

Continual performance evaluation and improvement is an essential ingredient for successful organisations, regardless of the type of the industry they belong to. It is the means to ascertain how well the organisation is achieving its desired business goals and then to decide pertinent strategies and tactics to improve operations. This exercise lends numerous benefits for organisations by way of:

- enhancing the profitability by reducing process costs and improving productivity;
- allowing them to identify their: (1) best and the most profitable operations and expand them, and (2) underperforming elements and improve them;
- permitting benchmarking of their performance against competitors.

At the core of performance evaluation and improvement is the acquisition of data related to various operations within the organisation, integrating them, analysing them and producing trends, patterns and forecasts to inform effective decision making in areas of production, resource management, marketing, etc. This process is also referred to as business analytics, which is the practice of iterative, methodical exploration of an organisation's data with the emphasis on discovering new knowledge and insights for optimising business processes. Traditionally, statistical methods were utilised for carrying out the analyses. However, the nature of data that is being produced by businesses has evolved tremendously with the high utilisation of the Internet of Things for business operations. The traditional statistical analysis techniques encounter significant practical difficulties in processing the new datasets. Among the key challenges are:

- **Scalability (volume)** – advancements in data generation and capturing methods have made dataset sizes of gigabytes, terabytes or even petabytes common nowadays, which were previously considered large.
- **High dimensionality** – it is common that datasets today have hundreds of attributes whereby a few decades ago datasets were described by a handful of attributes.
- **Complexity and variety** – traditional datasets contained attributes of the same type, either numerical or categorical. However, present day datasets

consist of webpages with semi-structured text, spatial data, time-series measurements, audios, videos, images, etc.

Hence, sophisticated approaches are needed for extracting useful information and knowledge from the non-traditional datasets. Data mining, also known as knowledge discovery (Kotu and Deshpande 2015), is an approach that deploys specialised computational methods to discover novel and useful patterns, trends, associations and knowledge in large datasets, which might otherwise remain unknown, to support important decision-making. It also enables building predictive models (Han *et al.* 2012; Tan *et al.* 2006; Kotu and Deshpande 2015). Data mining is a key enabler of analytics whose goal is to transform data/facts into actionable insights to influence decisions. In business contexts, therefore, data analytics is used as a synonym for data mining (knowledge discovery) (Delen 2015). Data science is another term used for data mining and analytics.

Types of data mining and analytics

Figure 1.1 illustrates the types of data mining and analytics tasks that may be performed on large datasets. These tasks fall under one of two broad categories; predictive data mining/analytics or descriptive data mining/analytics. Brief outlines of the different data mining tasks are provided below.

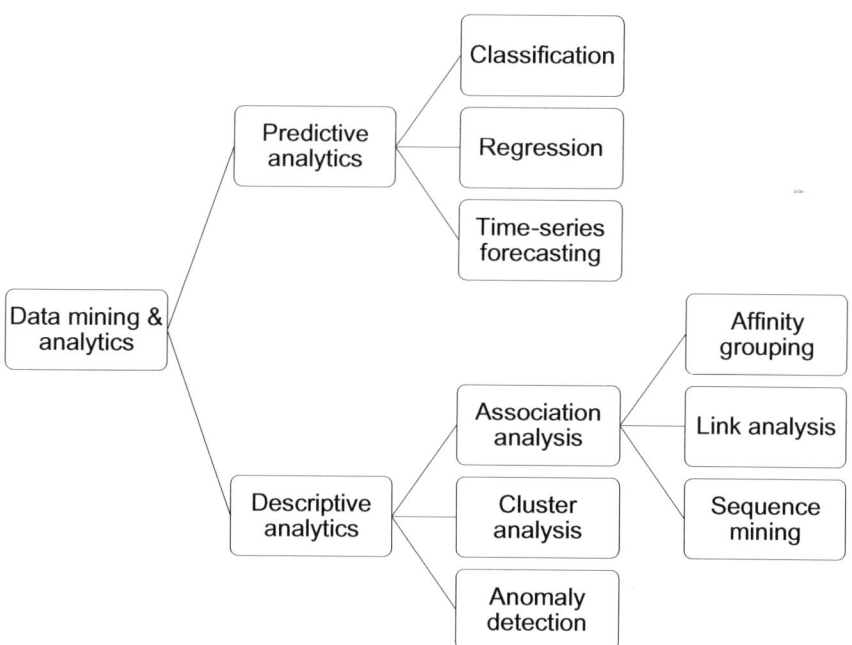

Figure 1.1 Types of data mining and analytics.

Predictive analytics

Predictive analytics produces models to predict future values of particular attributes (dependent or target variables) based on values of other attributes (independent or explanatory variables). It involves the task of building a generalised model from a previously known dataset for a target (dependent) variable as a function of explanatory (independent) variables for forecasting. Predictive analytics can take one of the three forms below, depending upon the nature of the output variable.

- **Classification** – is discovery of a predictive function for target variables that are categorical or polynomial. For instance, developing a model for making bid/no bid decisions based on some known project criteria is classification.
- **Regression** – refers to the discovery of a predictive function whose target variable is continuous. For example, a function for deciding the optimum mark-up percentage for a construction bid is a regression model.
- **Time-series forecasting** – a time series is a series of data associated with an equally spaced time interval (e.g. weekly, monthly, quarterly, etc.) about a business operation, which features a trend of variations over time. Data mining discovers a function that can predict future trends based on a current trend at a particular point-in-time. For instance, a building material manufacturer can use a time-series forecasting model to decide how much of a particular type of material to produce for the next three months based on the previous 12 months' orders.

Descriptive analytics

Descriptive analytics derives patterns (correlations, trends, clusters, trajectories and anomalies) that summarise subtle relationships in data. This is exploratory in nature and often requires further processing to explain the results. Association analysis, cluster analysis and anomaly detection fall under this category.

Association analysis

Association analysis refers to the derivation of association rules from historical data, which describe frequently occurring patterns in a business operation, for developing effective business tactics. In the context of the retail industry, this class of data mining is known as market-basket analysis. Three kinds of association patterns may be discovered in datasets:

(a) **Affinity grouping** (also known as frequent itemset mining) – affinity grouping is the task of determining which things go together. By mining the sale data shown in Table 1.2 for a grocery store, for instance, we may derive that customers who buy milk also tend to by diapers.

Rule: {milk} → {diaper}.

Table 1.2 Sales data

Sale Id	Purchases
1	Bread, Butter, Diaper, Milk
2	Eggs, Bread, Butter
3	Salmon, Diaper, Milk
4	Bread, Diaper, Salt, Milk
5	Tea, Eggs, Milk, Cookies, Diaper

Note
Retail chains can use affinity grouping to plan product placement on shelves or in catalogues and to design attractive packages. Online stores can use affinity grouping to identify cross-selling opportunities. For example, when you buy a book from an online bookstore such as Amzon.com, you may be shown a list of related books. Such lists are derived via affinity grouping. Similarly, mining of medical patient histories can indicate combinations of conditions that are heavily associated with increased risks of certain complications.

(b) **Sequence mining** (also known as frequent sub sequences) – while affinity grouping discovers things that happen together at the same time, sequence mining examines events in terms of their order of occurrence to discover association patterns over time. For example, sales data often contain temporal information about when and what items are purchased by customers. This information can be used to establish the sequence of purchases made by a customer over a certain period of time (Tan *et al.* 2006). By analysing all customers, sequence pattern rules are derived. Examples of such rules are:

- Customers who purchase a new laptop are likely to purchase a digital camera in the following six weeks, followed by a memory card in two weeks.
- When a bank customer asks for an account reconciliation, there is a good chance that he/she will close all accounts within the following four weeks.
- A construction worker who files a workers' compensation claim for electrocution is likely to file a claim for mental health issues in the next three months.

These rules are valuable for predicting future occurrences of events (Tan *et al.* 2006).

(c) **Link analysis** – links among objects may demonstrate certain natural patterns, which can provide useful insights for performance improvement. Link analysis discovers connections between objects/entities (e.g. organisations, people, documents, web pages, transactions, bank accounts, academic publications, etc.) and serves two primary purposes:

- Discovering patterns of interest between linked objects (e.g. social networks, business connections, disease epidemic paths, etc.),
- Discovering anomalies by detecting violated known patterns.

Link analysis has its origin in a branch of mathematics, called graph theory, and features visualisation strengths that can support better intuition and hypothesising about matters under study. It is therefore being utilised extensively in investigating criminal activities (e.g. counter terrorism, fraud detection) and in market research, medical research and search engine optimisation. In epidemiological studies, for instance, link analyses help establish/identify diseases with similar sets of contacts or transition patterns. Similarly, law enforcement authorities use link analysis to investigate crime scenes and to detect potential crimes by establishing connections between objects; e.g. people, background, weapon licences, addresses, vehicle registrations, phone calls, etc.

Cluster analysis

Cluster analysis refers to the process of classifying data into similar groups (clusters) so that objects within the same cluster are similar to one another and dissimilar to the objects in other groups/clusters. Some examples for the application of clustering in different fields are as follows:

- Marketing – marketing experts discovering distinct groups in their customer databases and then using this knowledge to develop targeted marketing programmes.
- Workplace safety – safety authorities identifying groups of workers with a high incident rate history and developing targeted incident prevention strategies.
- Medicine – a disease has several variations and each one is associated with a distinct group of risk factors. Clustering can be used to identify the pertinent group of risk factors that contribute to different disease variations.

Anomaly detection

Anomaly detection is the identification of data points, items or observations in the dataset that do not conform to the expected/normal behaviour pattern of a given group. Anomaly detection is also known as outlier detection and has a wide range of applications, including: fraud detection in credit card use and taxation, intrusion detection in cyber security, diagnosis of cancerous tumours and predictive maintenance of safety critical systems. In the building maintenance field, anomaly detection was used by Araya *et al.* (2017) to detect anomalous energy consumption patterns in buildings toward reducing energy waste.

Focus of the book

In light of the new direction and potentials identified above, this book aims at applying data mining and analytics techniques on past workplace incident data and discovering new knowledge and patterns that facilitate the development of innovative models and strategies to improve work health, safety and well-being in construction. Two broad objectives will be achieved in the chapters of the book, aligned with the central aim. These are:

- The methodological application of suitable data mining and analytics techniques to interrogate datasets for investigating different issues.
- The discovery of new knowledge and models related to the issues investigated to trigger pertinent prevention measures.

The accomplishment of the objectives will be demonstrated with data obtained in March 2016 from Safe Work Australia, which is an independent Australian government statutory body, primarily responsible for leading the development of policy to improve work health, safety and well-being, and workers' compensation arrangements across Australia. Amongst the functions it performs under this umbrella of responsibilities is to collect, analyse and publish data relating to WHS and workers' compensation in order to inform the development or improvement of policies related to these matters. Workers' compensation claims data collected by Safe Work Australia over a 13-year period, from 2002 to 2014, were obtained through a formal data request procedure. The obtained dataset consists of 391,494 cases of worker compensation claims filed by the construction industry across Australia during the 13 years.

Structure of the book

The book has six chapters and the contents of each chapter are outlined below.

Chapter 1: Introduction

This chapter establishes the key arguments and themes that propel the book and provides a context for the subsequent chapters. This chapter also provides a brief introduction to the field of data mining and analytics and sets the technical footings for the subsequent chapters.

Chapter 2: Curtailing construction fatalities using analytics

The construction industry accounts for a significant portion of workplace fatalities globally. Causes of fatalities do not exist in isolation; rather, associations of several factors create conducive circumstances for fatal incidents. The associations can be formed by factors related to workers, machinery and tools used, behaviour, work environment and management. It is important to discover such association patterns to create effective countermeasures for reducing fatal incidents. In-depth investigations of previous fatal incidents will help discover recurring associations that led to fatalities. However, commonly used statistical inferential methods and safety performance models cannot identify the combination of factors simultaneously. This chapter demonstrates the discovery of association patterns among construction workers, machinery, work activities and fatalities using data mining and analytics techniques for curtailing construction fatalities.

Chapter 3: Reducing uncertainties in compensation for occupational diseases in construction using analytics

Work-related diseases in construction have not gained enough attention from researchers although they are as critical as work injuries. These diseases impose significant physical and economic sufferings on workers who also encounter frustrating challenges in obtaining fair workers' compensation benefits. This is predominantly due to the difficulties in proving that work-related exposure to/ causes of diseases occurred in a particular project or employment, as there may be a time lag between the disease onset and the exposure. The difficulties imply uncertainties as to whether the disease would be compensable. Establishing a construction-industry-specific registry of association and causal patterns between occupational diseases and work-related factors would, to some degree, ease the difficulties inherent in proving work-related causes/exposure for obtaining compensation. This chapter, through data mining and analytics of past workers' compensation data, discovers relationships among occupational diseases, worker and work characteristics in the construction industry, to establish the said registry.

Chapter 4: Curbing psychological injuries in construction using analytics

The construction industry is notorious for high levels of work stress and related psychological injuries. Yet, psychological well-being of construction operatives and professionals is an underexplored subject in the construction domain as opposed to physical injuries and diseases. Psychological injury is a broad term that refers to any form of mental ill-health caused by work stress and includes: anxiety, depression, mood disorder, substance use, suicidality, etc. Because psychological risks suffered by construction operatives and professionals are invisible and silent, unlike physical injuries and diseases, they tend to go unnoticed for extended periods, causing serious damage to individuals, their families and society. Identifying work-related causes and their degree of impact on psychological health is fundamental for initiating effective preventive measures. This chapter demonstrates the utilisation of data mining and analytics techniques for understanding patterns of, and the risk factors for, psychological injuries in the construction industry.

Chapter 5: Predicting and preventing secondary psychological injuries in construction using analytics

There has been little research into secondary psychological injuries (SPIs) of construction workers. SPIs develop over a long period, following work-related physical injuries, due to workers experiencing negative changes to their socio-economic and work conditions, and quality of life, and poor injury claim experiences. Research outside the construction industry shows that SPIs pose significant long-term harmful consequences for workers, their families and employers. Nearly half of all injured workers develop subsequent psychological disorders and death from suicide is up to five times higher among people with SPIs than the general population and is most likely to occur within 5–6 years post injury. Construction being one of the most incident-prone industries, the risk of injured workers suffering SPIs is high. Yet, this is an unexplored topic in construction. This chapter proposes a neural network model for identifying operatives who are at risk of developing SPIs to enable pro-active prevention.

Chapter 6: Conclusion

This chapter integrates all the findings in previous chapters to propose a new, comprehensive causation model for workplace health, safety and well-being incidents in the construction industry. The chapter then outlines the practical implications of the proposed new causation model. The chapter finally discusses the methodological contributions made by this book and future research directions.

References

Araya, D.B., Grolinger, K., ElYamany, H.F., Capretz, M.A.M. and Bitsuamlak, G.T. (2017). An ensemble learning framework for anomaly detection in building energy consumption. *Electrical and Computer Engineering Publications*, 144(2007): 191–206.

Äyrämö, S., Pirtala, P., Kauttonen, J., Naveed, K. and Karkkainen. T. (2009). Mining road traffic accidents. http://users.jyu.fi/~samiayr/pdf/mining_road_traffic_accidents. pdf (accessed 10 April 2016).

Beshah, T. and Hill, S. (2010). Mining road traffic accident data to improve safety: role of road-related factors on accident severity in Ethiopia. http://ai-d.org/pdfs/Beshah.pdf (accessed 5 April 2016).

Bureau of Labor Statistics (BLS) (2015). National census of fatal occupational injuries in 2014 (preliminary results). www.bls.gov/news.release/pdf/cfoi.pdf (accessed 18 April 2016).

Cheng, C., Leu, S., Cheng, Y., Wu, T. and Lin, C. (2012). Applying data mining techniques to explore factors contributing to occupational injuries in Taiwan's construction industry. *Accident Analysis and Prevention*, 48(2012): 214–222.

Chua, D.K.H. and Goh, Y.M. (2004). Incident causation model for improving feedback of safety knowledge. *Journal of Construction Engineering and Management*, 130(4): 542–551.

Chua, D.K.H. and Goh, Y.M. (2010). Case-based reasoning approach to construction safety risk assessment: adaptation and utilisation. *Journal of Construction Engineering and Management*, 136(2): 170–178.

Ciarapica, F.E. and Giacchetta, G. (2009). Classification and prediction of occupational injury risk using soft computing techniques: an Italian study. *Safety Science*, 47(1): 36–49.

Cigularov, K., Chen, P. and Rosecrance, J. (2010). The effects of error management climate and safety communication on safety: a multi-level study. *Accident Analysis and Prevention*, 42(5): 1498–1506.

Delen, D. (2015). *Real-world Data Mining: Applied Business Analytics and Decision Making*. Upper Saddle River, NJ: Pearson Education.

Dumrak, J., Mostafa, S., Kamardeen, I. and Rameezdeen, R. (2013). Factors associated with the severity of construction accidents: the case of South Australia. *Australasian Journal of Construction Economics and Building*, 13(4): 32–49.

European Union (2011). Socio-economic costs of accidents at work and work-related ill health. www.ilo.org/public/libdoc/igo/2011/468757.pdf (accessed 13 April 2016).

Gürcanli, G.E. and Müngen, U. (2013). Analysis of construction accidents in Turkey and responsible parties. *Industrial Health*, 51(6): 581–595.

Han, J., Kamber, M. and Pei, J. (2012). *Data Mining: Concepts and Techniques*, 3rd edn. Waltham, MA: Morgan Kaufmann.

Haslam, R.A., Hide, S.A., Gibb, A.G.F., Gyi, D.E., Pavitt, T., Atkinson, S. and Duff, A.R. (2005). Contributing factors in construction accidents. *Applied Ergonomics*, 36: 401–415.

Health and Safety Executive, UK (2015). Health and safety in construction sector in the Great Britain, 2014/15. www.hse.gov.uk/statistics/industry/construction/construction.pdf (accessed 12 April 2016).

Hsueh, S., Huang, C. and Tseng, C. (2013). Using data mining technology to explore labor safety strategy – a lesson from the construction industry. www.pakjs.com/journals/29(5)/29(5)9.pdf (accessed 14 April 2016).

Huang, X. and Hinze, J. (2003). Analysis of construction worker fall accidents. *Journal of Construction Engineering and Management*, 129(3): 262–271.

Jacinto, C. and Soares, C.G. (2008) The added value of the new ESAW/Eurostat variables in accident analysis in the mining and quarrying industry. *Journal of Safety Research*, 39: 631–644.

Karra, V.K. (2005). Analysis of non-fatal and fatal injury rates for mine operator and contractor employees and the influence of work location. *Journal of Safety Research*, 36(5): 413–421.

Kheni, N.A., Gibb, A.G.F. and Dainty, A.R.J. (2010). Health and safety management within small and medium-sized enterprises (SMEs) in developing countries: study of contextualised influences. *Journal of Construction Engineering and Management*, 136(10): 1104–1115.

Kotu, V. and Deshpande, B. (2015). *Predictive Analytics and Data Mining: Concepts and Practice with RapidMiner*. Waltham, MA: Morgan Kaufmann.

Kudyba, S. (2014). *Big Data, Mining and Analytics: Components of Strategic Decision Making*. Boca Raton, Florida: CRC Press.

Marques, P.H., Jesus, V., Olea, S.A., Vairinhos, V. and Jacinto, C. (2014). The effect of alcohol and drug testing at the workplace on individual's occupational accident risk. *Safety Science*, 68: 108–120.

Martin, L., Baena, L., Garach, L., Lopez, G. and Ona, J.D. (2014). Using data mining techniques to road safety improvement in Spanish roads. *Procedia – Social and Behavioural Sciences*, 160(2014): 607–614.

Murie, F. (2007). Building safety – an international perspective. *Building Safety*, 13(1): 5–11.

Panthi, K. and Ahmed, S.M. (2015). Predictive models from accident reports. *51st ASC Annual International Conference Proceedings*. http://ascpro0.ascweb.org/archives/cd/2015/paper/CPRT344002015.pdf (accessed 5 April 2016).

Pierson, L. (2017). *Data Science for Dummies*, 2nd edn. Hoboken, NJ: Wiley.

Rivas, T., Paz, M., Martín, J.E., Matías, J.M., García, J.F. and Taboada, J. (2011). Explaining and predicting workplace accidents using data mining techniques. *Reliability Engineering and System Safety*, 96: 739–747.

Safe Work Australia (2015a). Construction Industry Profile. www.safeworkaustralia.gov.au/sites/SWA/about/Publications/Documents/911/construction-industry-profile.pdf (accessed 13 April 2016).

Safe Work Australia (2015b). The Cost of Work-related Injury and Illness for Australian Employers, Workers and the Community: 2012–13. www.safeworkaustralia.gov.au/sites/SWA/about/Publications/Documents/940/cost-of-work-related-injury-and-disease-2012-13.docx.pdf (accessed 15 April 2016).

Silva, J.F. and Jacinto, C. (2012). Finding occupational accident patterns in the extractive industry using a systematic data mining approach. *Reliability Engineering and System Safety*, 108: 108–122.

Tan, P., Steinbach, M., and Kumar, V. (2006). *Introduction to Data Mining*, 1st edn. Boston: Pearson Addison Wesley.

Thipparat, T. (2012). Application of adaptive neuro fuzzy inference system in supply chain management evaluation. http://cdn.intechopen.com/pdfs/34230.pdf (29 March 2016).

Ural, S. and Demirkol, S. (2008). Evaluation of occupational safety and health in surface mines. *Safety Science*, 46(6): 1016–1024.

Waehrer, G.M., Dong, X.S., Miller, T., Haile, E. and Men, Y. (2007). Costs of occupational injuries in construction in the United States. *Accident Analysis and Prevention*, 39(6): 1258–1266.

Yoon, S.J., Lin, H.K., Chen, G., Yi, S., Choi, J. and Rui, Z. (2013). Effect of occupational health and safety management system on work-related accident rate and differences of occupational health and safety management system awareness between managers in South Korea's construction industry. *Safety and Health at Work*, 4(4): 201–209.

2 Curtailing construction fatalities using analytics

Introduction

The construction industry provides about 7 per cent of global employment but accounts for a highly disproportionate 40 per cent of workplace fatalities worldwide (Murie 2007). Moreover, of all industrial sectors, construction records one of the highest number of fatalities. For example, in the UK, 43 fatalities were recorded for the construction industry during 2015/16 while fatalities in other industries were lower; services – 37, agriculture – 27 and manufacturing – 27 (HSE 2017). In Australia, the construction fatality rate was 3.2 per 100,000 workers in 2015, double the rate of all industry fatalities in the same period (1.6 per 100,000 workers) (Safe Work Australia 2016).

Construction fatalities impose significant costs and burden on victims' families, employers, and the overall economy. Matthews *et al.* (2011) reported that, apart from losing loved ones, victims' families experience numerous negative long-term consequences on their physical and mental health, social life, and economic conditions. Likewise, fatalities cause a range of commercial impacts on businesses such as loss of production, damage to plant and equipment, lowered morale, adverse publicity, and loss of future business (CDC Group 2009). Construction injuries and fatalities cost the overall economy hundreds of millions of dollars every year due to lost economic outputs. Manzo IV (2015) estimated the annual economic costs of construction fatalities and injuries in three states of US, Illinois, Indiana and Iowa, and these were US$270 million per year, US$150 million per year and US$125 million per year, respectively. It is therefore of paramount importance to curtail construction fatalities.

Fatalities do not occur due to a single factor; rather, associations of several factors create circumstances conducive to fatal incidents. The associations can be formed by factors related to workers, work activities, machinery and tools used, and the mechanism of the incident. Understanding these association patterns around construction fatalities is fundamental for implementing pertinent prevention measures. To this end, this chapter aims to investigate:

- What association patterns recur among factors related to workers, work activities, machinery and tools used, and the mechanism of incidents in construction fatalities?

- What innovative measures might be implemented to curtail construction fatalities?

Methods and materials

The research primarily adopted data mining of past fatal incident records for the Australian construction industry. Discoveries of the data mining exercise were compared, contrasted and corroborated with existing literature on construction fatalities. The Multiple Correspondence Analysis (MCA) technique, which is an extension of Correspondence Analysis (CA), was selected for discovering the association patterns. CA and MCA belong to the family of exploratory analysis methods that reveal patterns in complex datasets. Specifically, MCA is capable of detecting and presenting underlying structures in large datasets, encompassing multiple categorical variables, and is particularly pertinent to studies that collect large amounts of qualitative data (Costa *et al.* 2013). Other advantages of this technique are:

- It does not require any distributional assumption; i.e. the dataset is not required to satisfy a normal distribution.
- It provides insights into the dataset with information visualisation; i.e. graphical representations of underlying structures in datasets are produced.

Roux and Rouanet (2010) pointed out that although MCA is a powerful tool for pattern recognition and visual presentation, it is still under-utilised in many promising fields, specifically in the domain of safety where very few studies are noted; for instance, Das and Sun (2015) analysed fatal vehicle crashes by MCA, and Maiti *et al.* (2014) mined safety rules for derailment in a steel plant using correspondence analysis. MCA is a new tool for construction safety research, with high potential. Discussions on the technical/mathematical details of CA and MCA are beyond the scope of this book. Readers who are interested to know the procedure of performing correspondence analysis and the method of explaining the results are suggested to refer to the case example under the Help menu of IBM SPSS.

Data

Construction fatality data required for the research was obtained in March 2016 from Safe Work Australia, which is a government agency responsible for leading the development of national policy to improve work health and safety. It compiles work-related compensation claims and fatality datasets of Australian states and territories across all industries. When a formal request for construction incident and compensation data was made to the federal body by the author, they collated data from the different chapters and provided them to the author upon formally signing a confidentiality agreement. The entire process took about 12 months, but the dataset was large, encompassing 391,494 cases of workers' compensation claims filed by the construction industry across Australia over a

Table 2.1 Attribute definition

Attribute label	Attribute definition
Date of accident	The date the injury occurred, or the occupational disease was first reported to the employer
Date of reporting	The date the claim was reported to the employer
Date of claim	The date the claim was lodged with the insurer by the employer
Date determined	The date the insurer accepted or denied liability for the claim
Work status	The claimant's last known work status
Industry of employer	Main activity of the employer at the time of reporting the incident
Size of employer	The number of full-time workers employed by the enterprise for which the claimant works
Year of birth	Year of birth of the worker making the claim
Gender	The gender of the worker
Occupation	The worker's occupation at the time of the injury/disease
Duty status	The worker's duty status at the time of the injury/disease
Hours usually worked each week	The number of hours and minutes usually worked each week (including overtime) by the injured worker
Normal weekly earnings (in $)	Gross weekly earning rounded in whole dollars
Industry of workplace	Main activity of the employer at which the worker was injured or experienced the exposure resulting in disease
Postcode of workplace	Australian postcode
Nature of injury/disease	The most serious injury/disease sustained or suffered by the worker
Bodily location of injury/disease	Part of the body affected by the most serious injury/disease
Mechanism of injury/disease	The action, exposure or event that was the direct cause of the most serious injury/disease
Agency of injury/disease	The object, substance or circumstance directly involved in inflicting the most serious injury/disease
Breakdown agency of injury/disease	The circumstance (i.e. hazardous work practices or environment) that principally led to the most serious injury/disease
Severity outcomes	Severity of the work-related injury/disease (fatality, permanent incapacity or temporary incapacity)
Time lost (hours)	The number of hours and minutes lost for which compensation was paid
Compensation ($)	All payments made to the worker or worker's family in compensation for the death, injury or disease

Table 2.2 Selected variables for data mining

Variable	Measurement	Classification system followed
Age	Under 20; 20 to 29; 30 to 39; 40 to 49; 50 to 59; 60 & above	Derived from ABS classifications
Gender	Male; Female	Natural
Occupation	Managers; Professionals; Technicians and trade workers; Community and personal service workers; Clerical and administrative workers; Sales workers; Machinery operators and drivers; Labourers	ANZSCO classification
Nature of injury	Intracranial injuries; Fractures; Amputation, internal organ damage, wound and lacerations; Burn/electrocution; Nerves and spinal cord injuries; Trauma to joints/ligaments/muscles/tendons; Other injuries; Musculoskeletal and connective tissue diseases; Mental disorders; Digestive system diseases; Skin disease; Nerves system disease; Respiratory system disease; Circulatory system disease; Infectious disease; Cancer; Other diseases	TOOCS3.0 classification
Bodily location affected	Head; Neck; Trunk; Upper limbs; Lower limbs; Multiple locations; Systemic locations; Psychological system (non-physical location); Unspecified location	TOOCS3.0 classification
Mechanism of incident	Falls; Hitting objects with a part of the body; Being hit by moving objects; Sound and pressure; Body stressing; Heat, electricity and other environmental factors; Chemicals and other substances; Biological factors; Mental stress; Vehicle accident	TOOCS3.0 classification
Agency of incident	Machinery and fixed plant; Mobile plant and transport; Powered equipment, tools and appliances; Non-powered equipment, hand tools and appliances; Chemicals; Materials and substances; Environmental agencies; Animal, human and biological agencies; Other agencies	TOOCS3.0 classification

13-year period between 2002 and 2014. Filtering the database, a subset of 1048 cases of approved fatality claims was extracted for this study. A typical case was characterised by 23 attributes and Table 2.1 explains the definitions of these attributes.

Data pre-processing

Data pre-processing was performed to: (1) remove noise; (2) handle records/fields with missing values; (3) filter relevant attributes for data mining; and (4) redefine attribute measurements to enable effective association mining with MCA. Upon a careful examination of the dataset and the nature of the attributes used to record cases, seven out of 23 attributes of the dataset were considered adequate for pattern discovery. The original measurement scales of the attributes were revised as shown in Table 2.2. All the attributes were made categorical to enable non-parametric data mining. The categories used for the attributes were derived from different classification systems that exist in Australia. The mechanisms of incident, for instance, were based on the Type Of Occurrences Classification System (TOOCS) Third Edition of the Australian Safety and Compensation Council. Similarly, occupation classifications were based on the Australian and New Zealand Standard Classification of Occupations (ANZSCO) First Edition. Grouping intervals for age was derived from the classifications used by the Australian Bureau of Statistics (ABS) in its various reports. Of the 1048 cases, one was missing several entries for the attributes, and was therefore removed from the dataset.

MCA results and discussions

IBM SPSS Version 24 was deployed for association discovery. MCA can be performed with multiple dimensions (axes); however, Greenacre and Blasius (2006) suggested that a two-dimensional representation is adequate to explain the majority of variances in the data. Moreover, Das and Sun (2015) suggested that morphological maps are an effective way of presenting information visually as they allow one to interpret the distribution of variable category combinations. Hence, the MCA analysis for this study was performed with two dimensions and the graphical results are presented below, which include: the model summary, discrimination measures, and the joint plot of category points.

The model summary, presented in Table 2.3, reveals that 86.2 per cent of the variabilities in the dataset is captured by the two-dimensional space, as explained by the total inertia. Moreover, both dimensions have eigenvalues significantly higher than 1.0, which is the threshold value for accepting a dimension. The Cronbach's α values for the dimensions are much higher than the minimum acceptable threshold value of 0.5, suggesting that the categories represented by the dimension have a significant, shared covariance and measure the same underlying concept. These figures indicate that the MCA model is robust for association pattern discovery.

Table 2.3 MCA model summary

Dimension	Cronbach's alpha	Variance accounted for	
		Total (eigenvalue)	Inertia
1	0.849	3.678	0.525
2	0.672	2.359	0.337
Total		6.037	0.862
Mean	0.780[a]	3.019	0.431

Note
a Mean Cronbach's alpha is based on the mean eigenvalue.

Figure 2.1 displays the discrimination measures produced for the dataset. The plot maps the data attributes along the dimensions to enable one to deduce an appropriate dimension name; an exercise similar to what is performed in principal component analysis. As shown in the figure, variables related to workers are arranged along dimension 1 (horizontal axis) whilst variables related to

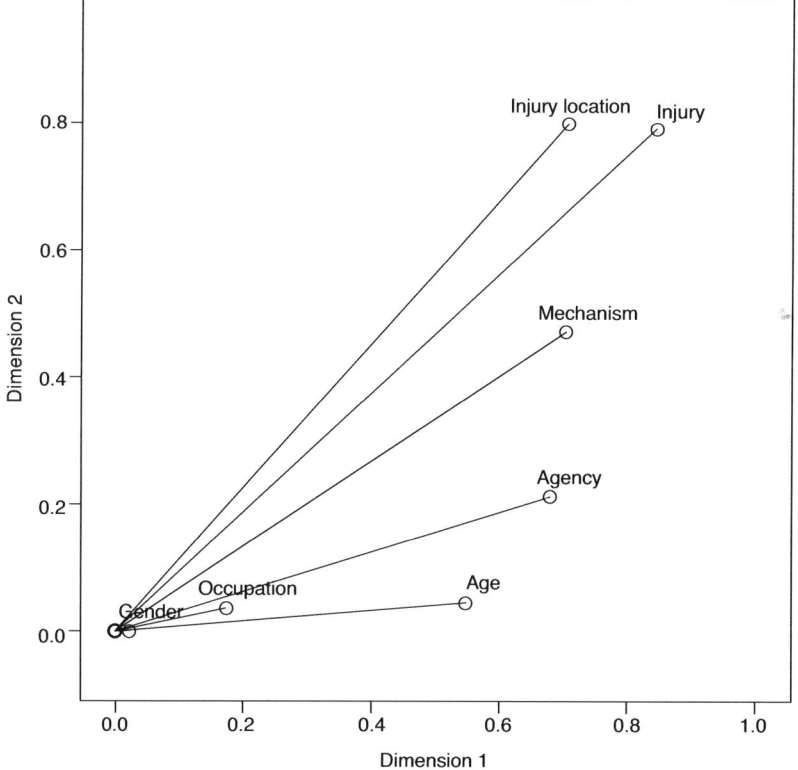

Figure 2.1 Discrimination measures.

incidents/injuries are arranged along dimension 2 (vertical axis). Hence, dimension 1 and dimension 2 could be labelled as worker characteristics and incident characteristics, respectively.

Figure 2.2 displays the most important output of the data mining exercise, which is the mapping of the association patterns. Fatalities in construction have been a concern for many WHS authorities and researchers alike around the world and, as such, many analyses and research have been undertaken by these authorities to understand the profile of construction fatalities. Through a review of the literature, it is understood that a notion of the 'Fatal Four' in construction has been established by these entities, which identify the fatal incident types. They are: falls, electrocutions, struck-by-object events, and caught-in/between-object events (e.g. Ling *et al.* 2009; Reece and Eidson 2006; Hinze and Russell 1995; Hinze *et al.* 1998). These events claim hundreds of lives every year in the construction industry and account for about 60 per cent of all construction incidents – including non-fatal ones. In the United States, for example, in 2015,

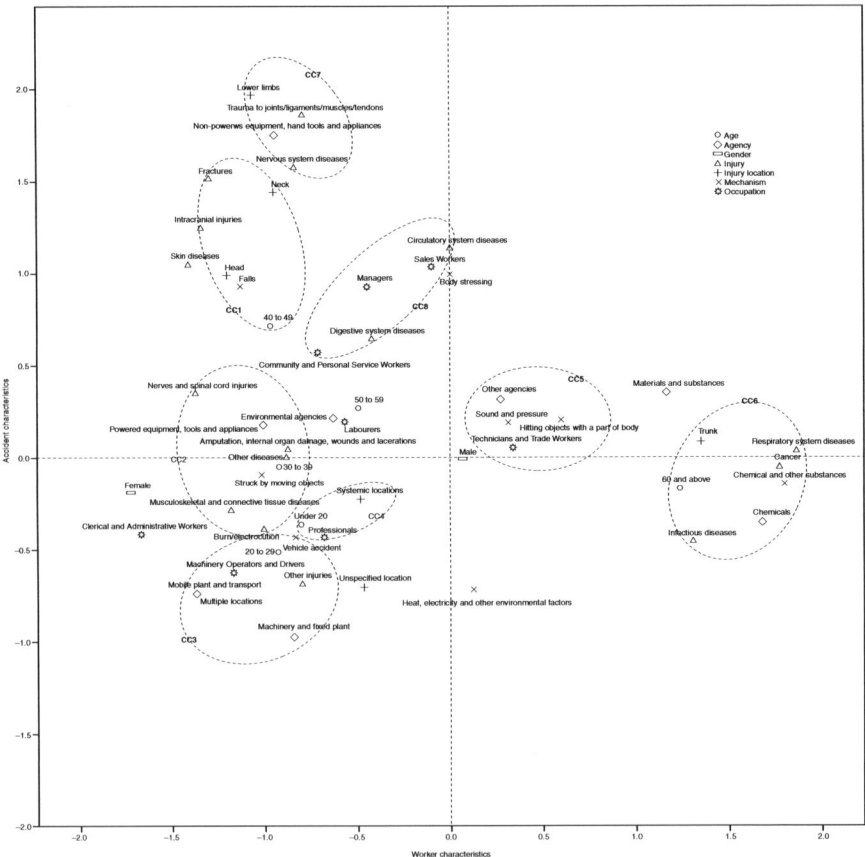

Figure 2.2 Joint category plot.

the principal causes of construction worker fatalities were falls, followed by struck-by object events, electrocution, and caught-in/between object events. These 'Fatal Four' accounted for more than half (64.2 per cent) of reported construction worker fatalities. Another example is Northern Ireland, where over the period 2015 (fourth quarter) to 2017 (first quarter) there were a total of five workplace fatalities recorded and all in the construction industry – three caused by 'falls' and two were due to struck-by incidents. Nonetheless, the discovery of this data mining exercise reveals more fatal injury patterns in the Australian context. In Figure 2.2, attribute values that co-occurred in construction fatalities are clustered together. Combination clouds can be drawn encircling these clusters, and insights into recurring association patterns in construction fatalities could be derived. Accordingly, seven combination clouds are drawn in Figure 2.2, covering key fatal incident patterns in the Australian construction industry, which are:

CC1 – Fatal falls
CC2 – Fatalities by powered tools
CC3 – Plant and machinery related fatalities
CC4 – Fatalities from mental stress
CC5 – Noise-induced fatalities
CC6 – Fatalities due to contact with chemicals and substances
CC7 – Fatalities by non-powered tools

The following sections discuss these in detail.

Fatal falls

Work-related slips, trips and falls result in a considerable number of injuries and fatalities reported globally. Despite the magnitude of the issue having been recognised and many attempts made to address it over several decades, it still persists as a major cause of work-related injuries and fatalities. Whether a fall of few metres or few storeys, a workplace fall can claim a life in seconds. Any worker, regardless of the extent of experience, can be vulnerable to a fall if proper safety measures are not in place. In the USA, for example, despite the Occupational Safety and Health Administration (OSHA) strictly requiring the implementation of adequate fall protection for work at a height of 1.8 m or above, the violation of this rule has been frequently reported as the prime cause for the prevalence of fall-related fatalities (Monforton *et al.* 2016). In 2015, there were 648 fatal work injuries in total for falls to a lower level, this was a decrease of 2 per cent from the count for 2014. Out of the 538 cases with a known height of fall, more than 40 per cent were falls of 4.5 m or lower and about 20 per cent involved falls from more than 9 m. Only around 9 per cent of all fatal falls occurred from a height of less than 1.8 m (Bureau of Labor Statistics 2016). This is consistent with the figures available from the UK where in the period 2014/2015, 35 workers were killed in the construction industry, with

15 of these being from slips, trips and falls from a height (HSE 2015). In Korea, between 1997 and 2004, falls accounted for 52.7 per cent of the 10,276 fatal occupational injuries recorded (Im *et al.* 2009).

Scaffoldings, ladders, elevated platforms, and roofs have been identified as agents that are responsible for a high proportion of fall-related fatalities. In the USA in 2015, for example, scaffoldings were the primary source of fatalities for 58 workers, ladders were involved in 90 deaths, and roofs were the primary source of 109 deaths. Fall protection, scaffolding, and ladders were in the top ten most cited violations of the Federal OSHA in the fiscal year 2016 (Jones 2017). This is comparable to Australia where, in 2015, 26 construction workers died due to falls from a height. Of these, six were from roofs, three were from ladders, two from scaffoldings and two from elevated work platforms (Safe Work Australia 2016). A study conducted at Loughborough University, UK in 2005 found that one reason for the multitude of issues relating to scaffoldings comes from the associated strike hazards created by the way scaffoldings are assembled; i.e. poor configuration, site constraints or equipment limitations (Haslam *et al.* 2005). Similar figures are seen in the Gaza Strip where 48 per cent of all workplace fatalities are from the construction industry. Of these, the majority of incidents were caused by individuals falling while working on roofs, scaffolds or ladders (Enshassi and Mohammaden 2012). The highest percentage of fatalities in the construction industry, 17 per cent of the total deaths, came from those working in roofing activities (Jensenius 2017).

Most of the existing literature relates fatalities to the height of the fall and/or the agents of incidents. Combination Cloud 1 (CC1) in Figure 2.2 provides a different perspective to fall-related fatalities in the Australian construction industry. Falls are generally one of the most common incident types in construction, yet not every fall results in a fatality. According to CC1, falls that resulted in serious damage to head or neck as well the ones that involved workers aged between 40 and 49 led to fatalities. There is a commonly believed proposition that older workers possess greater knowledge, skills, experience and patience on the job than their younger counterpart, and there is a documented negative correlation between age and incident rates (Frone 1998; Stalnaker 1998). However, when incidents happen, older workers are generally more severely wounded, and fatalities are more frequent among them (Janicak 2008; Siu *et al.* 2003; Stalnaker 1998). Table 2.4 shows the number of compensation claims, categorised by the age group, which were filed for falls from heights in the Australian construction industry in 2013 (Safe Work Australia 2013). According to APA (2016), over a 12-year period (2000/2001 to 2012/2013), older employees in all occupation categories in the Australian construction industry had higher rates of serious claims than younger workers. The percentage of approved serious claims for employees aged 55 and above doubled in 11 years, from 9 per cent in 2000–2001 to 18 per cent in 2011–2012. Since the global population is aging rapidly and the retirement age has been raised in some countries to address labour shortages, it is expected that the workforce in the near future will comprise a larger proportion of older employees (Siu *et al.* 2003). Hence, considerations of the impact of age on safety should be revitalised.

Table 2.4 Serious claims due to falls from height, 2010–2011

Age group (years)	Number of claims	% of claims	Incidence rate	Median time lost (weeks)
Younger than 25	1120	14	0.6	4.2
25–34	1410	18	0.6	6.0
35–44	1750	23	0.8	6.2
45–54	1935	25	0.9	7.0
55 and over	1510	20	1.0	7.8
Total	7730	100	0.7	6.2

Source: Safe Work Australia (2013, p. 21).

Fatalities by powered tools

Combination Cloud 2 (CC2) shows that workers who handle powered tools are exposed to struck-by objects, and electrocution/burn incidents that lead to death. OSHA (2002) categorised powered tools used in construction as electric tools, pneumatic tools (powered by compressed air), hydraulic power tools, liquid fuel tools, and powder actuated tools. Among the vast number available under these categories, Anderson (2009) and Graziano (2017) identified the nine most dangerous powered tools used in construction, namely: power nailers/ nail guns, chain saws, table saws, circular saws, power drills, air compressors, jackhammers, wood chippers and sanders.

The primary hazard associated with electric-powered tools is an electric shock that causes heart failures and burns, leading to death. An electric shock can also trigger falls from an elevated platform or ladder, leading to death or serious injuries (OSHA 2002). CDC (2007) reported that nail guns sent 22,200 workers to the emergency department during 2001–2005 in the US. Common injuries are: being shot by a nail from a gun or being struck by a nail gun, resulting in death, or musculoskeletal injury, fractured bones, eye injuries or puncture wounds. Fatalities were caused when the nail entered through the chest, neck or the head of the worker. Fatalities happened in similar ways with powered drills, whereby the spinning bit entered through the worker's chest, neck or the head (Anderson 2009). Powered saws are represented in struck-by or cutting fatalities, which occur when the operating blade bounces back into the worker's chest, a broken blade flies and cuts the worker or the worker loses balance or hold of the machine when working at a height and slashes him/herself. Fuel-powered tools and air compressors cause fatalities by explosions.

CC2 also argues that amputations, internal organ damage, wounds and lacerations as well as nerves and spinal cord injuries are quite prevalent among construction workers who use powered tools. Here, it is the age group of 30 to 39 that is more vulnerable to this form of fatal injuries. It was explained by CPWR (2017) that long-term use of powered tools can damage blood vessels and nerves in hands and fingers. Moreover, vibration produced when operating tools such as rotary hammers, chainsaws, grinders, and jackhammers travels through the hand, causing Raynaud's Syndrome (CPWR 2017). Bovenzi and Hulshof (1999) proved that occupations with an exposure to whole-body vibration (WBV) increase the risk for low back pain (LBP) disorders. Vitharana and Chinda (2017) found that a short-term exposure to WBV can cause increased heart rate, hyperventilation, headache, loss of balance, motion sickness, muscle fatigue, effects on speech, and effects on vision, whilst long-term exposure can cause degenerative disorders of spine, spine disc diseases and failures, lower back pain (LBP), and gastrointestinal system disorders.

Plant and machinery related fatalities

Combination Cloud 3 (CC3) reveals the types of fatal injuries that machinery operators and drivers on construction sites suffer, which are predominantly vehicular (run over) incidents and burn/electrocutions. The age category of 20 to 29 is at higher risk of fatalities and, intuitively, mobile plant, machinery, and fixed plant are the responsible agents. Moreover, the injuries damage multiple locations of the body. Further explorations of this finding are undertaken below with reference to other literature and reports.

Previous researchers listed numerous mechanisms of plant- and machinery-related fatalities in construction, including: run over by plant, struck-by moving plant/parts/materials carried/lifted, plant overturning, fall from plant, electrocution, crushed between plant and other objects, and engulfment/entangled/caught in running machinery (Pratt *et al.* 2001; McCann 2006; Beavers *et al.* 2006). In the USA, according to the OSHA, the Bureau of Labor Statistics, 10.1 per cent of all construction worker fatalities in 2013 were related to struck-by incidents, and 75 per cent of these involved vehicles and large equipment, such as cranes. Moreover, 67 out of 937 total deaths in construction in 2015 (7.2 per cent) were caused by getting caught in or between objects or equipment, a significant rise compared with 39 worker deaths in 2014. Likewise, in Australia, 8 per cent of the total construction fatalities between 2003 and 2013 were due to being trapped in or between equipment (Safe Work Australia 2015).

Plant-related fatalities in heavy and civil engineering construction largely involve run over incidents, when workers in this industry sector work alongside heavy machinery to construct roads, bridges, power plants and dams. Gameng (2016) reported that the heavy and civil engineering construction sector recorded the highest number of fatalities from vehicle incidents (18 fatalities) between 2003 and 2014 in Australia. The situation is much more severe in the US where 268 construction workers suffered death in road construction projects between 2003 and 2010, of which 100 workers (37 per cent) were killed by dump trucks and 131 by a reversing vehicle or mobile plant. The vulnerable occupation type is not limited to workers alone but includes truck drivers, plant operators, foremen and operating engineers (CPWR 2013).

Several mobile plant and machinery are used on construction sites, such as augers, backhoes/excavators, cranes or derricks, concrete pumps, trucks, rollers/compactors, forklifts, cherry pickers, front-end loaders, and bulldozers/graders.

Lingard *et al.* (2013) found that excavators/backhoes, trucks, and cranes were more frequently involved in plant- and machinery-related fatalities on construction sites. Additionally, cranes were represented largely in struck-by incidents and trucks were in run over incidents, whilst excavators were involved in fatal incidents of run over-by, struck-by, overturning and engulfment.

Lingard *et al.* (2013), qualitatively analysing coroner reports, identified 12 causes of plant- and machinery-related fatalities, which are: failure to adequately segregate plant and people at the worksite; unsafe actions of plant operators; communication failures between the plant operator and the signal personnel; poor lighting and visibility in the plant operational area; site constraints; ineffective supervision at the worksite; unsafe plant design (lack of reversing cameras, alarms, sensors, etc.); workers' knowledge and skills in safe operations of and around plant; poor plant maintenance regimes; unsafe construction process design; poor risk assessment practices; and the lack of coordination between different trades operating at the same time in the same area. Efforts to minimise plant- and machinery-related fatalities in construction require addressing these causes.

Fatalities from mental stress

Combination Cloud 4 (CC4) concerns workers aged under 20 and professionals. The primary reason for fatalities among these cohorts is mental stress. The mental health of construction workers and professionals is believed to be affected by poor psychosocial working conditions such as high job demand, low job autonomy, lack of support and lack of job security (Gullestrup *et al.* 2011; Heller *et al.* 2007). There is a lot of evidence to suggest that suicide is a worrying cause of fatalities in the construction industry, and poor mental health is the trigger of it. Doran *et al.* (2015, p. 2) calculated that in 2012, a total of 169 construction workers committed suicide and for every suicide, there were 15 attempts with three (17 per cent) resulting in full incapacity and 12 (83 per cent) resulting in short absence. This situation persisted for over two decades, with Gullestrup *et al.* (2011) reporting higher rates of suicide for Australian construction workers than the general male population, between 1990 and 2006. These statistics are mirrored elsewhere in the world. For example, it was reported in the UK that one construction worker commits suicide every two days – a higher rate than any other industrial sector (Pearson and Broughton 2003).

Apprentices and young workers aged between 15 and 24 were 10 times more likely to commit suicide than die from a work incident (Stewart 2014). The higher rate in younger construction workers may be due to the pressure associated with a 'masculine' industry, which has been reported to have a bullying culture, mostly directed toward apprentices and young workers (Heller *et al.* 2007). This is made worse by the fact that males are less likely to seek support than females to militate against stress caused by workplace bullying (Olafsson and Jóhannsdóttir 2004). Workplace bullying experienced by blue-collar workers has been found to predict depression (Kivimäki *et al.* 2003; Agervold and Mikkelsen 2004) and has been linked to suicide in the UK and Norway (Rayner *et al.* 2002).

The higher rate of suicide in the construction industry may be linked to work–home conflicts too (Amagasa *et al.* 2005). It has been reported that construction workers and professionals commonly work up to 80 hours in a six-day work week to meet the work demands. The long work days not only impact on family time but also on quality of life, with reduced involvement in social and recreational activities, thus affecting psychological health (Heller *et al.* 2007). Long work days and weeks can contribute to marital dissatisfactions, a strain on relationships and families, and may lead to separation or divorce, which has been recognised as a risk factor for suicide, particularly among males (Kposowa 2000; Amagasa *et al.* 2005). Furthermore, psychological autopsy investigations revealed that construction workers who committed suicide were more likely to have separated/divorced or have serious relationship issues preceding the death. The construction work environment itself is stressful and the loss of a spouse through separation may exacerbate the already stressful work life and diminish social support networks while increasing financial strain (Heller *et al.* 2007).

Noise-induced fatalities

Combination Cloud 5 (CC5) indicates an interesting association between sound and hitting objects with a body part as the mechanism for fatalities among tradespersons and technicians in construction. Exposure to noise at work may cause many physiological and psychological responses, which become risk factors for fatalities.

Constant exposure to loud noise at construction sites is a direct cause of noise-induced hearing loss for workers. Workplace-noise induced hearing loss is an irreversible condition that can affect a person's life greatly. Long-term exposure to loud noise makes the nerve receptors in the inner ear die irreparably (ACT 2017). A 10-year follow-up longitudinal study conducted in the US found that 30 per cent of workers whose longest job was in construction trades reported fair or poor hearing (CPWR 2013).

Noise creates safety hazards for workers, interfering with communication and making warnings and alarms harder to hear, which may result in workers hitting moving and lifted objects. Moreover, workers with an impaired hearing or reduced hearing ability due to working in construction may have difficulties in hearing warning signals and alarms from their surroundings even under normal noise levels, exposing them to hitting and struck-by incidents. Dineen (2001) reported that constant exposure to loud noise impacts on workers' ability to perceive hazards and vigilance, which results in increased incidents. High levels of noise can also create stress, leading to irritability and headache, fatigue and aggression, and psychiatric disorders (enHealth Council 2004).

Apart from the hearing loss, and psychological and safety hazards, continued exposure to loud noise can lead to serious cardiovascular diseases. The human body reacts to excessive noise with abnormal secretion of hormones such as adrenaline and cortisol. Prolonged high levels of these hormones may lead to serious health effects such as high blood pressure, accelerated heart rate, narrowing of blood vessels, hypertension, increasing the risk of strokes and heart attacks, and reduced white blood cell count and immune system responses (WorkCover Queensland 2017).

The types of workers who are vulnerable to noise-induced risk include operators of impact equipment and tools (e.g. piling hammers, concrete breakers, manual hammers), users of explosives (e.g. blasting, cartridge tools), operators of pneumatically powered equipment, operators of plant powered by internal combustion engines, service and equipment maintenance workers, workers near noisy plant, and operators and others in enclosed spaces that house noisy activities or operating machinery (Government of Western Australia, n.d.).

Fatalities by chemicals and substances

One increasingly dire issue facing workers in the construction industry is air pollution, which has been linked to both acute and chronic health effects such as heart diseases, respiratory diseases, cancer and allergic disorders. Combination Cloud 6 (CC6) reveals that chemicals and other substances are strongly associated with construction fatalities caused by cancer and respiratory system diseases in Australia. Workers aged 60 and above are predominantly represented in this type of workplace fatalities.

According to the World Health Organization, more than three million people die prematurely owing to prolonged exposure to air pollution. Occupational exposure to air pollution can be caused by a variety of factors, including worksite location and weather conditions. According to the London Atmospheric Emissions Inventory in the UK, construction sites are collectively responsible for emitting about 7.5 per cent of total damaging nitrogen oxides (NOx), 8 per cent of large particles (PM10) and 14.5 per cent of the most dangerous fine particles (PM2.5) (HSE 2010). Most of these emissions come from the multitude of heavy equipment operating on sites across the UK; e.g. generators, diesel diggers, cranes, and other such machines. In 2005, the Health and Safety Executive (UK) found that, on average, more than 230 construction workers die each year from cancer caused by exposure to diesel fumes. The occupation types that substantially contribute to these figures include construction personnel in general, metal workers, painters and decorators, repair trades, and roofers. Asbestos continues to be a leading cause of fatalities due to airborne pollutants in the construction industry, with 2717 fatalities recorded due to this in 2005 (HSE 2010). HSE (2010) stated that 56 per cent of cancer registrations in men were attributable to work in the construction industry (mainly mesotheliomas, lung, bladder and non-melanoma skin cancers). While there are 21 sectors that include 100 or more attributable registrations, this figure differs when focusing purely on the statistics of workplace fatalities. When fatalities alone are reviewed, it becomes clear that construction remains one of the leading sectors for fatal cancer in the UK (HSE 2010).

Fatalities by non-powered tools

Fatalities due to using hand tools in construction are relatively low though injuries are common. A wide range of non-powered tools are used in construction, such as: striking tools (hammer, sledge); cutting tools (knife, axe, handsaw, chisel); digging tools (shovel, pickaxe, spade, hoe, trowel); prying tools (crowbar); smoothing tools (plane, file); and tightening and loosening tools (screwdriver, wrench, clamp). They all have the potential to cause serious injuries from misuse and improper maintenance (Lingard 2004) and different types of injuries are suffered by workers, including: cuts, scrapes and punctures; hit by falling tools; trip and fall on tools; falls due to losing balance when using hand tools at heights; struck by objects removed/cut by hand tools; and body stressing. These are confirmed by Combination Cloud 7 (CC7) that shows the injury types and incident mechanisms associated with non-powered tools. Nerve system diseases and trauma to joints/ligaments/ muscles/tendons appear to be the main injuries suffered by these workers. Non-powered tools are also equally distanced from incident mechanisms such as falls and body stressing in the joint plot, suggesting they can be present in falls and body stressing incidents as agents. The use of non-powered hand tools in construction can contribute to musculoskeletal disorders too (Kumar 2014). These are caused by wear and tear in the body and injuries to joints, bones and nerves, affecting hands and wrists, the shoulder, the neck and upper back, the lower back, and the hips and knees (CPWR 2017; Vitharana and Chinda 2017).

Age and fatalities

Age is an often-debated factor not only in workplace fatalities but also as a general factor in employment. There has been an abundance of research over the last few decades looking at the effects of age on workplace injuries. Despite the multitude of research, there does not seem to be a consensus around whether younger or older workers are more likely to be the victims of workplace incidents.

On one side of the argument, research indicates that the fatality rate for young workers was lower than the overall rate (for example, Driscoll *et al.* 2001; Feyer *et al.* 2001; Chen and Fosbroke 1998; Ruser 1998; Bailer *et al.* 1998). This was evident in the USA where older workers are at higher risk of workplace fatalities. For instance, in 2015:

- Fatality rates were generally lower among younger workers, aged 25 to 34 years (2.3 per 100,000 full-time workers);
- Workers aged 45 years and older represented 58 per cent of work fatalities while accounting for only 45 per cent of the total hours worked;
- Thirty-five per cent of all fatalities occurred in workers aged 55 or older (1681 deaths); and

- Workers aged 65 years or older were more than 2.5 times more likely to die on the job than other workers, with a fatality rate of 9.4 per 100,000 full-time workers.

(AFL-CIO 2017; Bureau of Labor Statistics 2016)

On the other hand, another set of research argues that the fatality rate for young workers was higher than the overall rate (for example, Thelin 2002; Jackson and Loomis 2002; Collins *et al.* 1999; Rabi *et al.* 1998). Research conducted in the UK into the relationship between the age and experience of operatives and their level of safety performance showed a strong link between the two variables. The results indicated that operatives aged between 16 and 20 were more likely to fall victim to incidents than were other age groups. Further analyses of the data revealed that the rate of incidents tends to decline steadily after the age of 28 and reaches the lowest in the mid-40s. This phenomenon was clarified by Sawacha *et al.* (1999) and Cherns (1966) as workers continually adapting and acquiring skills and rapporteur with age, which enable older workers to perform risky and highly demanding tasks more safely.

This research endeavoured to clarify this contradiction. The bar graphs illustrated in Figure 2.3 show the spread of fatalities across different age groups in

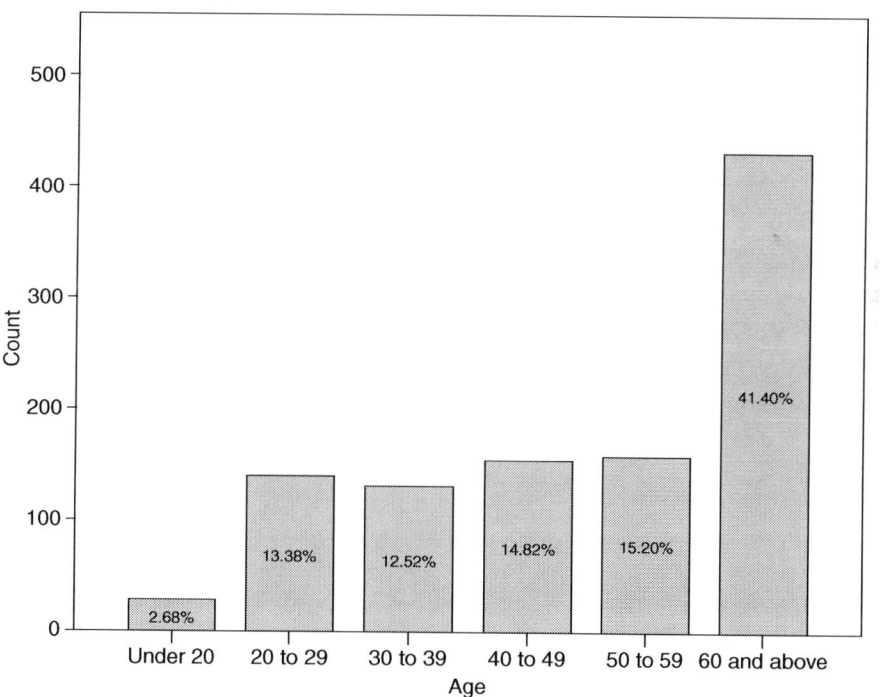

Figure 2.3 Age and fatality.

the dataset investigated. Fatalities among workers aged 60 and above are almost three times the other age groups, except for workers under 20 who account for less than 3 per cent. Moreover, age groups such as 20 to 29, 30 to 39, 40 to 49 and 50 to 59 are very similar. These figures reconcile the contradiction present in the existing literature and statistics that the percentage of workers who suffer from fatal workplace injuries is consistently around the 15 per cent mark until the age group of 60 and above, where there is a significant increase in the percentage of fatalities. Furthermore, considerations of the association patterns in the joint plot reveal that: (1) fatalities due to falls are predominant among workers aged 40 to 49; (2) labourers aged 50 to 59 are more susceptible to fatal incidents than other age groups: (3) workers aged 30 to 39 are mainly exposed to struck by moving objects, resulting in fatalities; (4) workers aged below 20 and 20 to 29 are susceptible to vehicular incidents, leading to fatalities; and (5) workers above 60 are vulnerable to pollution-related fatalities but it does not show the occupation type.

Interference of nature with construction fatalities

The preceding sections described work factors associated with construction fatalities. Natural phenomena have been responsible for construction fatalities too. Construction activities are heavily influenced by and exposed to weather conditions, which can amplify existing health and safety hazards on construction sites, leading to new and sometimes unanticipated risks for workers. Natural agents of construction fatalities are discussed in the following sections.

Lightning

Lightning is a 'struck-by' element that has had an impact on construction. Construction workers who undertake activities outdoors in open spaces, on or near tall structures, or near conductive metals are significantly exposed to lightning risks. Examples of construction activities at high lightning risk are heavy equipment operation, roofing, scaffolding, building maintenance, and plumbing and pipe fitting. In the US, for instance, between 1995 and 2002, there were 374 deaths by lightning in the general population with 129 of these being work-related incidents. Of the 129, 30 per cent were involved in construction-related activities, and 34 per cent were in agriculture and fishing (Adekoya et al. 2005). Over a period of 10 years, from 2006 to 2016, 352 people were killed by lightning strikes in the United States, a slight decrease from the previous study mentioned. Of this new total, 31 per cent were farm workers, and 30 per cent were involved in construction activities. Overall, 30 per cent of the deaths caused by lightning strikes involved construction workers on the job.

Seasonal low temperature – snow/ice

Weather that is usually linked to the winter season, such as snow, sleet and ice, significantly increases the hazards present in construction activities. For example, snow is a major hazard in electrical situations as the extreme moisture reduces the insulation properties of protective equipment. In addition, the potential for incidents working at height, on equipment, and within work sites is also increased due to the reduced friction and increased lubrication on smooth surfaces. Falls are the leading cause of worker fatalities and injuries during these weather conditions, with workers being more prone to falling off or through roofs, and from ladders and aerial lifts. Less common, but no less concerning, are the injuries and fatalities caused by roof collapses due to the added weight of layers of snow and ice. Proper use of work zone traffic controls is even more important during wet or snowy conditions, as drivers are more likely to skid or lose control of their vehicles in poor weather conditions, such as those created by snow and ice. Reaction and correction times are much slower in wet conditions, especially in those with the additional complication of ice.

Heat stress

Conversely, construction workers who are exposed to extreme heat or work in adversely hot environments/conditions may be at risk of heat stress, causing occupational illnesses, injuries and in some situations, fatalities (Xiang *et al.* 2014; Jay and Kenny 2010). This is a more substantial issue for those who work in countries where extremes of weather occur more frequently. Xiang *et al.* (2014) noted changes to incident rates on construction sites after an estimated threshold temperature of 37.7°C. When the weather reaches extreme highs, workers should maintain hydration, ensure protection from the sun's strong rays and take a break more frequently (Navon and Kolton 2006). Such self-regulation can be an effective method to manage heat stress; however, concerns over productivity loss can prevent that from being implemented on construction sites. Since sub-contracting is the norm rather than the exception, whereby the payment is based on the progress achieved (Rowlinson *et al.* 2014; Miller *et al.* 2011), self-regulation is hard to achieve. Not only does self-management, therefore, become impossible from a financial point of view, but it is very difficult in view of the hectic schedules presented by construction projects. The consequence is a substantial increase in incidents (Navon and Kolton 2006). This is confirmed by the findings of an analysis of mortality due to heat stress during the period 2000 to 2010 in the USA (Gubernot *et al.* 2015), which revealed that construction workers were extremely vulnerable to heat stress, producing the second highest risk rate among all industries during that period. Similarly, in South Australia, the research of Rameezdeen and Elmualim (2017) concluded that construction workers were found to be at high risk of injury due to heat stress as workers are often outdoors, undertaking strenuous manual work.

Curtailing construction fatalities with advanced digital technologies

Safety in construction is managed in two distinct phases: Design-for-Safety (DfS)/Prevention-through-Design (PtD) in the design phase; and job site safety management in the construction phase.

DfS/PtD involves conducting systematic and detailed risk analyses in various design sub-phases to identify hazards and risks created by a proposed design, and introducing alternative design solutions that meet both the client's needs and local building codes while eliminating the hazards or reducing the risks for construction workers (Kamardeen 2015). Nonetheless, not every hazard can be removed by design substitutions. Site safety management, on the other hand, entails controlling the residual hazards that could not be eliminated by design changes as well as other project-specific hazards through engineering and administrative controls and making sure the construction work proceeds without harming people on site. This process comprises three primary iterative tasks: safety planning and implementation on site, continual safety monitoring and control/inspection, and safety training and induction of workers.

Recent developments in digital technologies have the potential to assist construction teams in improving these processes and thereby curtailing construction incidents. Among the noteworthy technologies are: visualisation technologies, unmanned aerial vehicles and wearable technologies. These technologies can help enhance safety management in both design and construction phases. The following sections explain the various ways these technologies may be implemented in construction projects.

Visualisation technologies for enhancing construction safety

Visualisation technologies such as Building Information Modelling (BIM), 4D simulations, immersive virtual reality (gaming technology) and augmented reality can help create simulations and animations of construction processes and operations to allow project team members to visually interact with and assess job site conditions, detect safety risks and then introduce preventive measures before the actual construction can start on site. The technologies offer great potentials to improve construction safety through enhanced practices in design-for-safety, safety planning, and safety training.

Design-for-Safety (DfS)

The role of the design team has traditionally been to produce building designs that meet the client's requirements and the designer's architectural aspirations whilst complying with the local building codes. The final shape of the building and the materials specified have a direct influence on the selection of construction methods and operations. These, in turn, determine the setting and environments in which people work, alluding to the conclusion that the design team plays a central role in determining the safety or danger inherent in a project (Brace *et al.* 2009). The notion of design-for-safety is grounded on this background and is applied to eliminate or mitigate hazards potentially encountered by construction workers on site.

Every design creates a unique construction context, which dictates the level of hazards and danger present. The construction context includes five elements: site settings (site terrain condition, space and accessibility, road and traffic conditions and vicinity); task settings (location of the task and the temporary structure needed to build it); materials (type and nature of materials used); plant and equipment (type and nature of operations); and workers (type and skill levels). An incident can emanate from one or a combination of these contextual elements. The implementation of DfS requires the design team to thoroughly know and analyse the construction context and safety consequences of their design options and then make safe design choices for a project (Kamardeen 2015, p. 2). However, with few exceptions, such as those extensively involved in design–build contracts, designers lack the skills and expertise that are a cornerstone for implementing DfS (Zhou *et al.* 2012) and they often rely on external safety consultants for this job. DfS consultants together with the design team or project teams in design–build contracts conduct charrette-style workshops to improve the design for safety, in which the safety experts' experience and imagination play a critical role. The construction contexts are visualised in the mind and dangers inherent are explained to the designers. The use of visualisation technologies can improve the dialogue and outcomes of this exercise tremendously (Sacks *et al.* 2015).

The construction process of a proposed design can be simulated in a virtual environment using a combination of BIM, 4D simulation and game engine platforms. The design team can be taken on virtual reality walkthroughs whereby hazards and safety consequence created by their design choices can be visually explained, persuading them to make changes to their design to eliminate danger to workers. Additionally, the immersive experience would facilitate for designers the learning and awareness of construction safety, which they would not normally have exposure to otherwise. This method would also enable designers to easily prove undertaking effective DfS analyses for their projects for legal purposes, if required.

Safety planning

Construction safety planning involves the contractor identifying hazards in a construction project and preparing detailed risk control methods to avert the danger to workers by those hazards. Traditionally, contractors derive suitable safety measures for a project by reading 2D drawings and hazard checklists, and the process heavily relies on the experience of the safety planner and produces static documents, namely safe work method statements. Hence, in the traditional approach, the link between the safety plan and the actual task execution, which is dynamic, is weak, leading to error-prone results (Azhar 2017). The

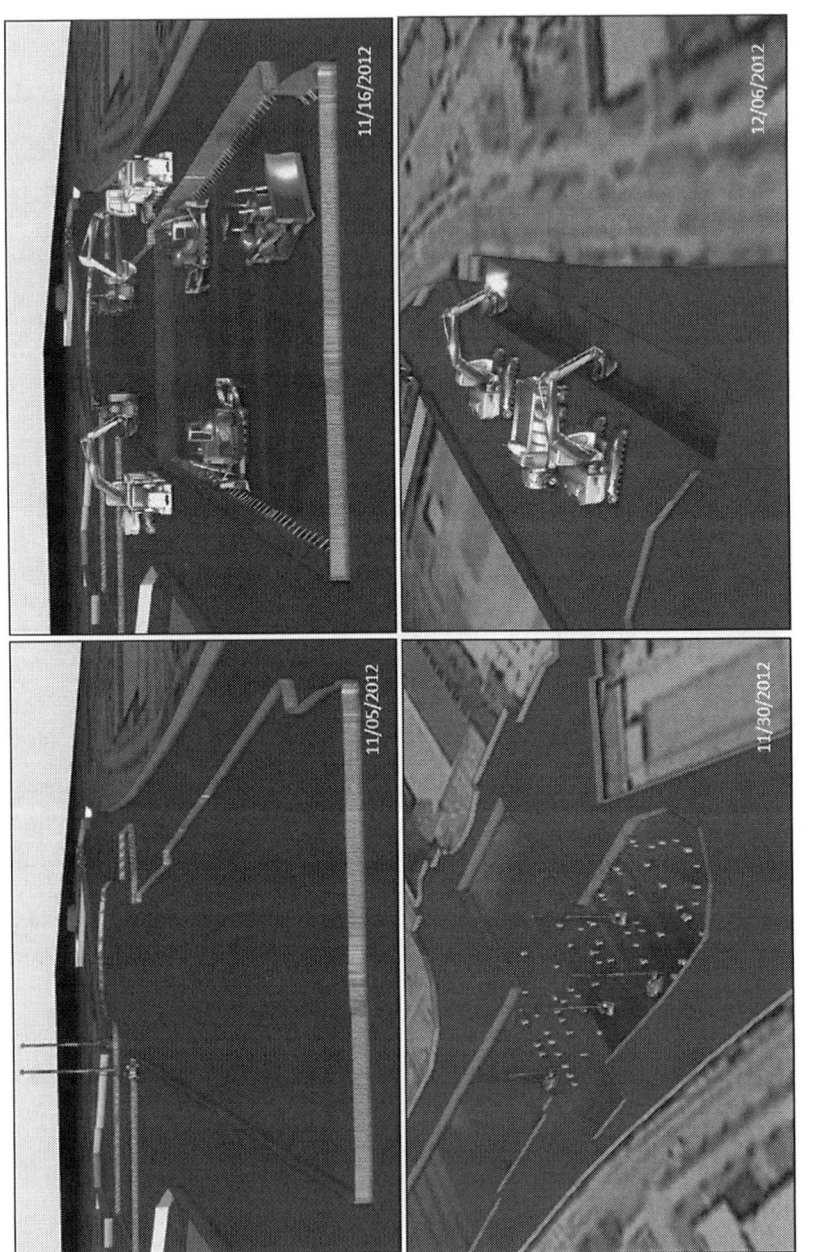

Figure 2.4 Visualisation-based safety plans.

Source: images courtesy of Salman Azhar.

Figure 2.4 Continued

utilisation of a virtual construction site, virtual walkthroughs and 4D phasing simulations would allow contractors to visually assess job site conditions and recognise hazards, resulting in more contextualised site layout and safety plans that enhance safety communications with the project team members and sub-contractors, and thereby enhance effective risk control on site (Azhar *et al.* 2012; Sulvankivi *et al.* 2012).

Azhar (2017) demonstrated the application of BIM and 4D simulation technologies for safety planning at the Recreation and Wellness Centre project at Auburn University, USA, specifically aiming to address the 'fatal four' incidents in construction. The demonstration case developed four safety plans: (1) excavation risk management plan to safely coordinate earth work operations at the job site; (2) crane operation risk management plan to identify the swing radius of the crane to ensure safe distance from power lines and nearby temporary and permanent structures; (3) fall protection plan for leading edges and roofers to identify multiple fall hazards that were not identifiable in 2D plan views and to introduce fall protection railings; and (4) an emergency response plan to decide effective emergency exits, routes and safe shelter locations for workers and equipment. They used software such as Autodesk Revit for BIM modelling, Google Sketchup for creating 3D equipment and characters for animations, Synchro for 4D phasing simulation and MS Project for scheduling. Figure 2.4 illustrates screenshots of their hazard analysis and safety planning. The main contractor's team involved in the project indicated that the new approach was moderately-to-highly accurate in hazard identification and was better in incident control on site compared with the traditional approach because: (1) it facilitated collabora-tive safety planning by engaging designers, engineers, the head contractor, and subcontractors in the planning phase; (2) it improved communications of the safety plan to the construction personnel, OSHA, and the owner; and (3) it enabled fully addressing the logistical details of safety tasks at the preconstruction phase, leaving the construction phase to focus fully on implementation aspects.

Safety training

Research has established that unsafe behaviour of workers has repeatedly caused a significant portion of construction fatalities. Choudhry *et al.* (2008) argued that workers practise unsafe behaviour due to inadequate safety know-ledge and awareness. The construction industry has been deploying several methods to train workers in safety knowledge and awareness, including: on-site safety toolbox talks, specialised safety training, apprenticeship pro-grammes and behaviour-based safety training (Kamardeen 2013). A variety of techniques and media, such as lecturers, presentations, seminars, discussions, video clips and case studies, are utilised to disseminate knowledge about site hazards and risks, and safe practices. Le *et al.* (2015) argued that workers in these training methods play a passive role, hence finding the programmes less engaging and quite boring, which impedes effective learning. Le *et al.* (2015) further suggested that experiential safety training, aided by Virtual Reality (VR) technologies, would be a better alternative to address these challenges. VR technologies enable the building of 3D immersive environments that can be explored by workers in real-time. VR modules can be developed to provide interactive, experiential learning opportunities for construction workers. These have been utilised extensively for training in high-risk occupations, namely pilots, military personnel, surgeons and nuclear power-plant operators. Davis (2016) suggested three ways that VR could be used to improve construction safety training:

- Construction sites and operations can be simulated in a virtual environment using a combination of advanced 4D modelling and game engine platforms. With VR headsets, workers can be teleported and immersed into the job site where they can virtual-realistically witness incidents; for example, the tipping over of a crane, collapse of an excavation, and falling off a ladder. Being immersed in such dangerous environments would nurture in the workers' minds safety consciousness and would encourage them to adopt safe behaviour and make careful decisions.
- Previous incidents can be recreated virtually to explain to workers the unsafe behaviour and other errors that caused the incident and subsequently teach them safe practices to prevent similar incidents occurring again.
- VR simulations of a construction project can be created prior to the actual construction commencing. With the simulated site, workers can be navigated through their potential tasks and made to point out all the hazards and hidden dangers facing them and then be shown the safe method of performing these tasks. Figure 2.5 shows an example of such a simulated site. The VR tool will improve safety toolbox talks that happen just before the start of an operation, in order to minimise incidents.

Figure 2.5. VR simulation for safety training.
Source: adapted from Davis 2016.

Unmanned aerial vehicles for improving safety monitoring and control

Regular safety inspections are one of the tenets of safety management, and serve two purposes: ensuring that the safety plans are implemented properly; and identifying the emergence of new hazards. In this process safety managers frequently walk around the job site and check and obtain real-time data through direct observations of the current site condition, construction activities, materials, equipment operations and workers' behaviour based on safety criteria, and interact with workers (Toole 2002). This process presents few challenges. For effective incident control, the safety inspection should occur daily. However, when the project is large and complex, and only one safety manager is available, the task becomes complex and less frequent. Moreover, safety managers should go to all areas of the construction site, ranging from the bottom of an excavation to the top of the roof structure and spend a short-time to obtain data. This is dangerous for safety managers and they may miss important data because of the short observation time in any area. These challenges may undermine the outcome of safety inspections, affecting their ultimate goal of incident prevention.

Small unmanned aerial vehicles, also known as 'drones', have the potential to address the challenges inherent in safety inspection and enhance the process outcomes. Drones can move faster than humans, can reach inaccessible areas of job sites and can be piloted remotely using a smartphone, tablet or a computer. Since they are equipped with video cameras and communication hardware to transfer real-time data, safety managers are able to obtain images and real-time videos from a range of locations in quick fashion. Even in long highway projects that extend for several kilometres, safety inspections can be done in a short time (Gheisari and Esmaeili 2016; Irizarry *et al.* 2012).

Wearable technologies for enhancing construction safety

Wearable technologies have been used extensively across industries such as healthcare, manufacturing, mining and athletics, to name a few, to improve performance. However, the applications of these technologies in construction are still at the nascent stage. Some examples of wearable technologies relevant to construction include Radio-Frequency Identification Devices (RFIDs), real-time location systems, smart helmets, smart safety vests, smart glasses, smart watches, wristbands and bionic suits. These technologies have ample potential to increase safety in construction, and some application areas are discussed below.

Real-time proximity detection and risk alert

In circumstances where workers must carry out tasks in the proximity of moving plant, RFIDs can be used to prevent struck-by incidents. RFIDs use radio waves of different frequencies to identify objects, and comprise two parts; and RFID tag and RFID reader. An RFID reader can be installed on the plant and workers in the plant operating zone would be provided with RFID tags that are detectable by the RFID reader. During the movement of the plant and workers, the RFID reader can be set up to automatically read the tag and alert the plant operator, the worker and the work supervisor of a possible collision if their proximity is too close (Lingard *et al.* 2013).

Lu *et al.* (2011) suggested further applications of RFIDs for safety precautions in other areas of a construction site. The site may be divided into zones, demarcated by different colours to indicate the level of hazards/dangers in the zone (e.g. green, amber and red). All the potential risks in a zone can be registered in RFID tags, which are ubiquitous on site. Workers will be equipped with RFID readers and an alarm system to detect the hazards automatically from the tags and alert workers real-time of potential danger when they enter a zone. The information can be transmitted to the appropriate line manager/supervisor for attention.

Monitoring PPE use

Not wearing appropriate Personal Protective Equipment (PPE) correctly has been identified as one of the causes of incidents in construction. Kritzler *et al.* (2015) proposed a system that uses wireless technology and wearable devices to ensure PPE required for a specific task is worn by the worker. The system comprises beacons and a smartwatch (wearable device). A beacon is a small, lightweight device that constantly emits a wireless signal, which can be read by the smartwatch, containing a mobile application, which is worn by the worker. The beacons, each having its identifier, are attached to pieces of PPE worn by a worker for a task, e.g. helmets, safety goggles, gloves, etc. Moreover, another set of beacons are attached to different work zones on a job site, which keeps the registry of required PPE for the work zone. The smartwatch worn by the worker receives two kinds of the signal; the identity of PPE required for the task from the beacons available in the work zone, and the identity of PPE worn by the workers. Both lists are compared to ensure proper PPE is worn for the job. Any mismatch of identity will trigger a warning, alerting the worker and the supervisor. This system also ensures that every individual worker wears his/her own personally fitted PPE, eliminating the hazards from incorrect size PPE.

Environmental sensing

Most construction activities are performed outdoors where workers are exposed to weather and environmental factors such as temperature, humidity, wind, pressure, air quality, etc. Some activities take place in confined spaces and locations where the presence of toxic gases or the absence of oxygen is possible. Some other activities involve the use of hazardous materials such as chemicals, gases and solid materials. The hazards posed by these elements are not visible to the eye and therefore close monitoring must be in place to prevent dangers to workers. Automated sensing of these environmental and weather-related hazards and eliminating them is now possible with the utilisation of recently introduced smart sensors and wireless sensor networks that can measure a range of concerns including temperature, light, air quality, humidity, pressure, colour, gas leaks, and toxic gases such as carbon monoxide and hydrogen sulphide (Briand *et al.* 2011; Swan 2012; Sensirion 2017). With the aid of the sensors worn by construction workers, highly localised, real-time data on the presence of hazardous environmental and atmospheric conditions can be monitored and potential risks to lives can be minimised proactively.

Conclusion

The construction sector is responsible for 40 per cent of global workplace fatalities. Previous research established that four incident types such as falls, electrocution, struck-by object events, and caught in/between object events are heavily represented in construction fatalities. This study performed data mining of a large dataset from Safe Work Australia, without setting any hypothesis or pre-assumptions. The data mining exercise discovered some insightful association patterns among worker characteristics, incident characteristics and work activity characteristics, which can be classified under seven clusters, namely: fatal falls, fatalities by powered tools, plant and machinery related fatalities, fatalities by mental stress, noise-induced fatalities, fatalities by chemicals and substances, and fatalities by non-powered tools.

WHS authorities, researchers and construction organisations have been working relentlessly to curtail construction incidents, especially fatalities, by introducing new and continually improving existing legislation, safety management frameworks, and best practice guides. It is suggested that the power and capabilities offered by the recent developments in digital technologies should be leveraged to raise the current state of safety planning, monitoring, and training in construction to a higher level. Three pertinent technologies are visualisation technologies, unmanned aerial vehicles and wearable technologies. Utilising visualisation technologies to conduct design-for-safety workshops would enable design teams to visualise the risky design options and trigger robust prevention through design practices. Likewise, the technology, when used by builders for site hazard analysis, safety planning and safety training, would help address many long-confronted challenges caused by the dynamic complexities of construction sites, skill shortages and a migrant workforce. The use of unmanned aerial vehicles for site safety inspections would enhance the speed and effectiveness of the process. Wearable technologies can make remarkable improvements to the safety of workers who operate in unobvious locations and/or are in constant motion around various hazardous zones on site.

Despite the many potential safety benefits, the implementation of these advanced technologies may face resistance from industry, primarily due to the extra expenditures required. Nevertheless, construction clients and organisations should realise the long-term business benefits that can be reaped through zero-fatality construction projects, and deem the expenses as return-guaranteed investments.

References

ACT (2017). Hazardous substances [Online]. Access Canberra. Available: www.accesscanberra.act.gov.au/app/answers/detail/a_id/3555/~/work-safe-in-the-workplace-building-and-construction#!tabs-20 (accessed 28 September 2017).

Adekoya, N. and Nolte, K.B. (2005). Struck-by-lightning deaths in the United States. *Journal of Environmental Health*, 67(9): 45–50.

AFL-CIO Safety and Health Department (2017). *Death on the Job: The Toll of Neglect*, 26th edn. April 2017. https://aflcio.org/sites/default/files/2017-04/2017Death-on-the-Job_0.pdf (accessed 1 May 2017).

Agervold, M. and Mikkelsen, E.G. (2004). Relationships between bullying, psychosocial work environment and individual stress reactions. *Work & Stress*, 18: 336–351.

Amagasa, T., Nakayama, T., and Takahashi, Y. (2005). Karojisatsu in Japan: characteristics of 22 cases of work-related suicide. *Journal of Occupational Health*, 47(2): 157–164.

Anderson, M. (2009). *Slide Show: The 10 Most Dangerous Power Tools*. www.forbes.com/forbes/welcome/?toURL=www.forbes.com/2009/12/21/most-dangerous-tools-business-healthcare-tools_slide.html&refURL=www.google.com.au/&referrer=www.google.com.au/ (accessed 11 October 2017).

APA (2016). *Tradies National Health Month Health 'Snapshot'*. Tradies Health Australia: The Australian Physiotherapy Association.

Azhar, S. (2017). Role of visualisation technologies in safety planning and management at construction job sites. *Procedia Engineering*, 171: 215–226.

Azhar, S., Behringer, A., Sattineni, A. and Maqsood, T. (2012). BIM for facilitating construction safety planning and management at job sites. *Proceedings of the CIB W099 International Conference on 'Modelling and Building Health and Safety'*, 10–12 September 2012, Singapore.

Bailer, A.J., Stayner, L.T., Stout, N.A., Reed, L.D. and Gilbert, S.J. (1998). Trends in rates of occupational fatal injuries in the United States (1983–92). *Occupational and Environmental Medicine*, 55: 485–489.

Beavers, J.E., Moore, J.R., Rinehart, R. and Schriver, W.R. (2006). Crane-related fatalities in the construction industry. *Journal of Construction Engineering and Management*, 132(9): 901–910.

Bovenzi, M. and Hulshof, C.T.J. (1999). An updated review of epidemiologic studies on the relationship between exposure to whole-body vibration and low back pain (1986–1997). *International Archives of Occupational and Environmental Health*, 72: 351–365.

Brace, C., Gibb, A., Pendlebury, M. and Bust, P. (2009). *Health and Safety in the Construction Industry: Underlying Causes of Construction Fatal Accidents – External Research*. Secretary of State for Work and Pensions, Inquiry into the Underlying Causes of Construction Fatal Accidents. www.hse.gov.uk/construction/resources/phase2ext.pdf, London.

Briand, D., Oprea, A., Courbat, J., and Bâesan, N. (2011). Making environmental sensors on plastic foil. *Materials Today*, 14(9): 416–423.

Bureau of Labour Statistics (2016). News release: National Census of Fatal Occupational Injuries in 2015. US Government.

CDC (2007). Nail-gun injuries treated in emergency departments – United States, 2001–2005. www.cdc.gov/mmwr/preview/mmwrhtml/mm5614a2.htm (accessed 11 October 2017).

CDC Group (2009). Preventing fatalities and serious accidents. http://toolkit.cdcgroup.com/assets/uploads/CDC_Good_Practice-Preventing_Fatalities_and_Serious_Accidents.pdf (accessed 30 October 2017).

Chen, G.X. and Fosbroke, D.E. (1998). Work-related fatal-injury risk of construction workers by occupation and cause of death. *Human and Ecological Risk Assessment*, 4(6): 1371–1390.

Cherns, A. (1966). Accident at work. Cited in A.T. Welford (1976) *Society Problems and Methods of Study*. London, UK: Routledge and Kegan Paul.

Choudhry, M.R., Fang, D. and Ahmed, S.M. (2008). Safety management in construction: best practices in Hong Kong. *Journal of Professional Issues in Engineering Education and Practice*, 134(1): 20–32.

Collins, J.W., Landen, D.D., Kisner, S.M., Johnston, J.J., Chin, S.F. and Kennedy, R.D. (1999). Fatal occupational injuries associated with forklifts, United States, 1980–1994. *American Journal of Industrial Medicine*, 36: 504–512.

Costa, P.S., Santos, N.C., Cunha, P., Cotter, J. and Sousa, N. (2013). The use of multiple correspondence analysis to explore associations between categories of qualitative variables in healthy ageing. *Journal of Aging Research*, Article ID 302163: 1–12.

CPWR (2013). The construction chart book, U.S. CPWR – The Center for Construction Research and Training.

CPWR (2017). Preventing musculoskeletal disorders in construction workers [Online]. Australia: Occupational Safety and Health Administration (OSHA). Available: http://elcosh.org/document/1648/d000560/preventing-muskuloskeletal-disorders-in-construction-workers.html (accessed 26 September 2017).

Das, S. and Sun, X. (2015). Association knowledge for fatal run-off-road crashes by multiple correspondence analysis. *IATSS Research*, 39(2): 146–155.

Davis, L. (2016). Three ways virtual reality could improve safety training. www.agcmass.org/blog/constructing-observations-the-agc-blog-297/post/three-ways-virtual-reality-could-improve-safety-trainings-3071 (accessed 14 October 2017).

Dineen, R. (2001). *Noise and Hearing in the Building Construction Industry: A Study on Workers' Views on Noise and Risk*. Video. Causes and Prevention of Hearing Loss. Colloquium (NAL), 23 October 2001.

Doran, C.M., Ling, R. and Milner, A. (2015). The economic cost of suicide and non-fatal behaviour in the Australian construction industry by state and territory. Available from: www.matesinconstruction.org.au/flux-content/mic-2013/pdf/Cost-of-suicide-in-construction-industry-final-report.pdf (accessed 20 February 2016).

Driscoll, T., Mitchell, R., Mandryk, J., Healey, S., Hendrie, L. and Hull, B. (2001). Work-related fatalities in Australia, 1989 to 1992: an overview. *Journal of Occupational Health and Safety, Australia and New Zealand*, 17(1): 45–66.

enHealth Council (2004). *The Health Effects of Environmental Noise – Other than Hearing Loss*. Canberra, Department of Health and Ageing.

Enshassi, A. and Mohammaden, A. (2012). Occupational deaths and injuries in the construction industry. *The Fourth International Engineering Conference – Towards Engineering of the 21st Century*.

Feyer, A.-M., Langley, J., Howard, M., Horsburgh, S., Wright, C., Alsop, J. and Cryer, C. (2001). The work-related fatal injury study: numbers, rates and trends of work-related fatal injuries in New Zealand 1985–1994. *New Zealand Medical Journal*, 114: 6–10.

Frone, M.R. (1998). Predictors of work injuries among employed adolescents. *Journal of Applied Psychology*, 83: 565–576.

Gameng, M. (2016). The fatal five construction injuries in Australia [Online]. Australia: Safety in the Construction Industry. https://blog.plantminer.com.au/fatal-five-construction-industries-in-australia (accessed 23 September 2017).

Gheisari, M. and Esmaeili, B. (2016). Unmanned Aerial Systems (UAS) for Construction Safety Applications. *Proceedings of Construction Research Congress*, 31 May–2 June 2016, San Juan, Puerto Rico, pp. 2642–2650.

Government of Western Australia (n.d.). Noise in construction: identification, assessment and control. www.commerce.wa.gov.au/sites/default/files/atoms/files/noise_newsletter_construction.pdf (accessed 12 October 2017).

Graziano, J. (2017). How power tools can lead to serious work injuries. www.vacomplaw.com/blog/2017/01/27/how-power-tools-can-lead-178229 (accessed 11 October 2017).

Greenacre, M. and Blasius, J. (2006). *Multiple Correspondence Analysis and Related Methods*. London: Chapman & Hall/CRC.

Gubernot, D.M., Anderson, G.B. and Hunting, K.L. (2015). Characterizing occupational heat-related mortality in the United States, 2000–2010: an analysis using the census of fatal occupational injuries database. *American Journal of Industrial Medicine*, 58(2): 203–211.

Gullestrup, J., Lequertier, B. and Martin, G. (2011). MATES in construction: impact of a multimodal, community-based program for suicide prevention in the construction industry. *International Journal of Environmental Research and Public Health*, 8: 4180–4196.

Haslam, R.A., Hide, S.A., Gibb, A.G.F., Gyi, D.E., Pavitt, T., Atkinson, S. and Duff, A.R. (2005). Contributing factors in construction accidents. *Applied Ergonomics*, 36: 401–415.

Health and Safety Executive (HSE) (2010). The burden of occupational cancer in Great Britain. Prepared by Imperial College London, the Institute of Environment and Health, the Health and Safety Laboratory and the Institute of Occupational Medicine. www.hse.gov.uk/research/rrpdf/rr800.pdf (accessed 2 August 2017).

Health and Safety Executive (HSE) (2015). Health and safety in construction sector in Great Britain, 2014/15. www.hse.gov.uk/statistics/industry/construction/construction.pdf (accessed 9 May 2017).

Health and Safety Executive (HSE) (2017). Fatal injuries arising from accidents at work in Great Britain 2017. www.hse.gov.uk/statistics/pdf/fatalinjuries.pdf (accessed 17 November 2017).

Heller, T.S., Hawgood, J.L. and Leo, D.D. (2007). Correlates of suicide in building industry workers. *Archives of Suicide Research*, 11(1): 105–117.

Hinze, J. and Russell, D.B. (1995). Analysis of fatalities recorded by OSHA. *ASCE Journal of Construction Engineering Management*, 121(2): 209–214.

Hinze, J., Pedersen, C. and Fredley, J. (1998). Identifying root causes of construction injuries. *ASCE Journal of Construction Engineering Management*, 124(1): 67–71.

Im, H.-J., Kwon, Y.-J., Kim, S.-G., Kim, Y.-K., Ju, Y.-S. and Lee, H.-P. (2009). The characteristics of fatal occupational injuries in Korea's construction industry, 1997–2004. *Safety Science*, 47: 1159–1162.

Irizarry, J., Gheisari, M. and Walker, B.N. (2012). Usability assessment of drone technology as safety inspection tools. *Journal of Information Technology in Construction (ITcon)*, 17: 194–212.

Jackson, S.A. and Loomis, D. (2002). Fatal occupational injuries in the North Carolina construction industry, 1978–1994. *Applied Occupational and Environmental Hygiene*, 17(1): 27–33.

Janicak, C.A. (2008). Occupational fatalities due to electrocutions in the construction industry. *Journal of Safety Research*, 39: 617–621.

Jay, O. and Kenny, G.P. (2010). Heat exposure in the Canadian workforce. *American Journal of Industrial Medicine*, 53: 842–853 [PubMed].

Jensenius, J.S. (2017). *A Detailed Analysis of Lightning Deaths in the United States from 2006 through 2016*. National Weather Service, NOAA. www.lightningsafety.noaa.gov/fatalities/analysis03-17.pdf (accessed 28 July 2017).

Jones, K. (2017). Combating worker deaths in the construction industry. www.constructconnect.com/blog/construction-safety/combating-worker-deaths-construction-industry/ (30 October 2017).

Kamardeen, I. (2013). *Electronic OHS Management Systems for Construction*. London: Routledge.

Kamardeen, I. (2015). *Fall Prevention through Design in Construction: The Benefits of Mobile Computing*. London: Routledge.

Kivimäki, M., Virtanen, M., Vartia, M., Elovainio, M., Vahtera, J. and Keltikangas-Järvinen, L. (2003). Workplace bullying and the risk of cardiovascular disease and depression. *Occupational and Environmental Medicine*, 60(10): 779–783.

Kposowa, A.J. (2000). Marital status and suicide in the National Longitudinal Mortality Study. *Journal of Epidemiology and Community Health*, 54(4): 254–261.

Kritzler, M., Tenfält, A., Bäckman, M. and Michahelles, F. (2015). Wearable technology as a solution for workplace safety. *Proceedings of the 14th International Conference on Mobile and Ubiquitous Multimedia*, 30 November–2 December, Austria.

Kumar, S. (2014). *Biomechanics in Ergonomics*. Boca Raton, Florida: CRC Press.

Le, Q.T., Pedro, A., Lim, C.R., Park, H.T., Park, C.S. and Kim, H.K. (2015). A framework for using mobile based virtual reality and augmented reality for experimental construction safety education. *International Journal of Engineering Education*, 31(3): 713–725.

Ling, F.Y.Y., Liu, M. and Woo, Y.C. (2009). Construction fatalities in Singapore. *International Journal of Project Management*, 27(7): 717–726.

Lingard, H. (2004). Occupational health and safety in construction project management. In: H. Lingard and S. Rowlinson (Eds), *Occupational Health and Safety in Construction Project Management*, 1st edn. London: Routledge.

Lingard, H., Cooke, T. and Gharaie, E. (2013). The how and why of plant–related fatalities in the Australian construction industry. *Engineering, Construction and Architectural Management*, 20(4): 365–380.

Lu, W., Huang, G.Q. and Li, H. (2011). Scenario for applying RFID technology in construction project management. *Automation in Construction*, 21: 101–106.

Maiti, J., Singh, A.K. Mandal, S. and Verma, A. (2014). Mining safety rules for derailments in a steel plant using correspondence analysis. *Safety Science*, 68(1): 24–33.

Manzo IV, F. (2015). The cost of construction injuries and fatalities in Illinois, Indiana and Iowa. https://illinoisepi.org/site/wp-content/themes/hollow/docs/wages-labor-standards/ILEPI-Economic-Commentary-Cost-of-Construction-Injuries-IL-IN-IA.pdf (accessed 30 October 2017).

Matthews, L.R., Bohle, P., Quinlan, M. and Rawlings-Way, O. (2011). Traumatic work-related death in the construction industry: experiences of victims' families. http://sydney.edu.au/health-sciences/research/work-health/workplace-death-report.pdf (accessed 11 July 2017).

McCann, M. (2006). Heavy equipment and truck–related deaths on excavation work sites. *Journal of Safety Research*, 37(5): 511–517.

Miller, V., Bates, G., Schneider, J.D. and Thomsen, J. (2011) Self-pacing as a protective mechanism against the effects of heat stress. *Annals of Occupational Hygiene*, 55: 548–555.

Monforton, C., Ojeda, M., Shelton, J. and Perez, C. (2016). *Worker Memorial Day Report 2016: Workplace Fatalities in the Houston Area*. www.coshnetwork.org/sites/default/files/uploads/Houston_WMD_Report_2016.pdf (accessed 2 May 2017).

Murie, F. (2007). Building safety – an international perspective. *International Journal of Occupational and Environmental Health*, 13: 5–11.

Navon, R. and Kolton, O. (2006). Model for automated monitoring of fall hazards in building construction. *Journal of Construction Engineering and Management*, 132: 733–740.

Occupational Safety and Health Administration (OSHA) (2002). *Hand and Power Tools*. www.osha.gov/Publications/osha3080.pdf (accessed 10 October 2017).

Olafsson, R. and Jóhannsdóttir, H.L. (2004). Coping with bullying in the workplace: the effect of gender, age and type of bullying. *British Journal of Guidance & Counselling*, 32(3): 319–333.

Pearson, A. and Broughton, T. (2003). The dark side of construction. www.building. co.uk/the-dark-side-of-construction/1029319.article (accessed 26 February 2016).

Pratt, S.G., Fosbroke, D.E. and Marsh, S.M. (2001). *Building Safer Highway Work Zone*. Cincinnati, OH: Department of Health and Human Services.

Rabi, A.Z., Jamous, L.W., AbuDhaise, B.A. and Alwash, R.H. (1998). Fatal occupational injuries in Jordan during the period 1980 through 1993. *Safety Science*, 28(3): 177–187.

Rameezdeen, R. and Elmualim, A. (2017). The impact of heat waves on occurrence and severity of construction accidents, *International Journal of Environmental Research and Public Health*, 14: 70.

Rayner, C., Howl, H. and Cooper, C.L. (2002). *Workplace Bullying: What We Know, Who is to Blame, and What Can We Do?* London: Taylor and Francis.

Reece, C.D. and Eidson, J.V. (2006). *Handbook of OSHA Construction Safety and Health*. 2nd edn. Boca Raton, FL: Taylor & Francis.

Roux, B.L. and Rouanet, H. (2010). *Multiple Correspondence Analysis*. Washington, DC: Sage.

Rowlinson, S., Yunyan Jia, A., Li, B., and Chuanjing Ju, C. (2014). Management of climatic heat stress risk in construction: a review of practices, methodologies, and future research. *Accident Analysis Prevention*, 66: 187–198.

Ruser, J.W. (1998). Denominator choice in the calculation of workplace fatality rates. *American Journal of Industrial Medicine*, 33: 151–156.

Sacks, R., Whyte, J., Swissa, D., Raviv, G., Zhou, W. and Shapira, A. (2015). Safety by design: dialogues between designers and builders using virtual reality. *Construction Management and Economics*, 33(1): 55–72.

Safe Work Australia (2013). Work-related injuries and fatalities involving a fall from height, Australia. www.safeworkaustralia.gov.au/system/files/documents/1702/falls-from-height.pdf (accessed 11 March 2019).

Safe Work Australia (2015). Work-related injuries and fatalities in construction 2003 to 2013. www.safeworkaustralia.gov.au/doc/work-related-injuries-and-fatalities-construction-2003-13 (accessed 11 March 2019).

Safe Work Australia (2016). Work-related traumatic injury fatalities Australia, 2015. www.safeworkaustralia.gov.au/system/files/documents/1702/work-related-traumatic-injury-fatalities.pdf (accessed 11 March 2019).

Sawacha, E., Naoum, S. and Fong, D. (1999). Factors affecting safety performance on construction sites. *International Journal of Project Management*, 17(5): 309–315.

Sensirion (2017). *Environmental Sensors*. Sensirion. www.sensirion.com/en/environ mental-sensors/ (accessed 27 October 2017).

Siu, O.L., Phillips, D.R. and Leung, T.-W. (2003). Age differences in safety attitudes and safety performance in Hong Kong construction workers. *Journal of Safety Research*, 34: 199–205.

Stalnaker, C.K. (1998). Safety of older workers in the 21st century. *Professional Safety*, 43: 28.

Stewart, J. (2014). Mates in construction, building industry suicide prevention program, aims to curb high suicide rate. www.abc.net.au/news/2014-10-07/mates-in-construciton:-suicide-prevention-program-gives-hope/5796814 (accessed 15 January 2016).

Sulvankivi, K., Teizer, J., Kivinieme, M., Eastman, C.M., Zhang, S. and Kim, K. (2012). Framework for integrating safety into Building Information Modelling. *Proceedings of the CIB W099 International Conference on Modelling and Building Health and Safety*, 10–12 September 2012, Singapore.

Swan, M. (2012). Sensor mania! The internet of things, wearable computing, objective metrics and the quantified self 2.0. *Journal of Sensor Actuator Net*, 1(3): 217–253.

Thelin, A. (2002). Fatal accidents in Swedish farming and forestry, 1988–1997. *Safety Science*, 40: 501–517.

Toole, T.M. (2002). Construction site safety roles. *Journal of Construction Engineering and Management*, 128(3): 203–210.

Vitharana, V.H.P. and Chinda, T. (2017). Structural equation modelling of lower back pain due to whole body vibration exposure in the construction industry. *International Journal of Occupational Safety and Ergonomics*, 25(2): 257–267.

WorkCover Queensland (2017). Effects of excessive noise. www.worksafe.qld.gov.au/injury-prevention-safety/hazardous-exposures/noise/effects-of-excessive-noise (accessed 12 October 2017).

Xiang, J., Bi, P., Pisaniello, D., Hansen, A. and Sullivan, T. (2014). Association between high temperature and work-related injuries in Adelaide, South Australia, 2001–2010. *Occupational Environmental Medicine*, 71: 246–252.

Zhou, W., Whyte, J. and Sacks, R. (2012). Construction safety and digital design: a review. *Automation in Construction*, 22: 102–111.

3 Reducing uncertainties in compensation for occupational diseases in construction using analytics

Introduction

Acute or chronic health problems that are caused or aggravated by work practices or conditions are referred to as work-related or occupational diseases (Safe Work Australia 2017). Riva *et al.* (2012) reported high prevalence of work-related diseases in the construction industry, with a peak among the elderly, but also significant occurrences among young people. They further asserted that the percentage of workers with limited fitness to work, caused by work-related diseases, is also high. Yet work-related diseases in construction have not gained enough attention from researchers even though they are as critical as work injuries. These diseases impose significant physical and economic suffering on workers who also encounter frustrating challenges in obtaining fair workers' compensation benefits. The workers' compensation system operates with the following primary objectives (Boden 1986):

- Complete coverage of injuries and illnesses arising out of and/or while on employment.
- A reasonable level of benefits, including full payment for medical and rehabilitation services.
- Prompt delivery of benefits.
- Effective delivery of benefits, i.e. a low expense-to-benefit ratio.

It is, however, necessary to prove that an injury or illness occurred 'out of and/or while on employment' to obtain compensation. It is generally easy for occupational injuries but proving occupational causalities for illnesses in construction is challenging due to the following uncertainties:

- Occupational diseases are usually not related to a unique/single occupational exposure. With construction jobs being project-based and transitional, it is difficult to prove that a disease was contracted in a particular project as it may take an extended period, beyond the project duration, for a disease to show symptoms.
- A disease can have several causes, both work-related and non-work-related. It may not be possible to isolate the contribution of each cause to the risk of developing the disease.

- A disease may develop years after an exposure or even after the exposure has ceased. In this case, establishing evidence about exposure to workplace hazard is difficult.
- Records of exposure to workplace hazards may not exist.

The difficulties in proving work-related exposure imply uncertainties about whether the disease would be compensable. Even if scientific evidence is produced, in the adversarial legal system the evidence may be interpreted differently at hearing, leading to profit-maximising insurers or self-insured employers controverting claims when the expected gain from controverting is greater than legal and administrative costs (Boden 1986). To reduce the negative experiences for workers with workers' compensation for occupational diseases, the uncertainty about whether the diseases are occupational in origin should be eliminated. This may be achieved in two ways:

- Maintaining an up-to-date registry of occupational diseases, with scientific evidence on the causal links with occupational exposures, for use in the workers' compensation process.
- Establishing statistical evidence showing the rates of incidence of varying diseases among different groups of workers exposed to specific work hazards.

Safe Work Australia (2015), through comprehensive reviews of scientific and medical evidence, produced a list of deemed diseases in Australia, which identifies occupations or exposures and associated diseases. This is an Australian version of the list of occupational diseases maintained by the International Labour Organization (ILO 2010), with additional information about associated occupations and risk factors. While the list satisfies the first suggestion above, it is not industry-specific enough to be helpful in minimising workers' compensation disputes in construction.. The second strategy suggested above could help identify industry and occupation specific diseases and risk factors. Godderis *et al.* (2015) proposed that performing data mining and analytics on past incident records can help discover diseases specific to workers from a specific industry sector and their risk factors. To that end, this chapter aims to discover relationships among occupational diseases, severities, and worker and work characteristics in the construction industry through analytics of past workers' compensation data.

Method and material

In a broader perspective, the study applies data mining and analytics on past disease incident records for the construction industry in Australia to achieve its aim. Discoveries from the data mining and analytics exercise are then compared, contrasted and corroborated with existing literature on occupational diseases among construction operatives. The specific details of the data utilised, and the analytic techniques applied are described below.

Data

Data required for the research were obtained in March 2016 from Safe Work Australia, which is a government agency responsible for leading the development of national policy to improve work health and safety, as described in the previous chapter. Filtering the database of 391,494 cases of worker compensation claims filed by the construction industry across Australia over a 13-year period between 2002 and 2014, a subset of 43,685 cases of occupational disease claims for construction operatives was extracted for this study. The occupational diseases have been recorded under nine broad categories, as follows:

1   Musculoskeletal and connective tissue diseases
2   Digestive system diseases
3   Skin and subcutaneous tissue diseases
4   Nervous system and sensor organ diseases
5   Respiratory system diseases
6   Circulatory system diseases (cardiovascular diseases)
7   Infectious and parasite diseases
8   Neoplasms (cancer)
9   Mental diseases

This chapter concerns only physical diseases covered by the first eight categories above; mental diseases are discussed separately in the next two chapters.

Data pre-processing

A typical case was characterised by 23 attributes as explained in Table 2.1 in Chapter 2. Data pre-processing was performed to: (1) filter relevant attributes for data analytics; (2) redefine attribute measurements to enable effective analytics; (3) handle records/fields with missing values; and (4) identify diseases that have bigger health impact for further analytics.

Upon a careful examination of the dataset and the nature of the attributes used to record cases, nine out of 23 attributes of the dataset were considered adequate for data mining and analytics. The selected attributes are: size of the employer, age of the operative, gender, occupation, mechanism of disease, nature of disease, location of disease, agency of disease and disease severity. Furthermore, the original measurement scales of the selected attributes were revised, and numerical attributes were made categorical to enable non-parametric data mining, as shown in Table 3.1. The categories used for the attributes were derived from different classification systems that exist in Australia. The nature of disease, mechanisms of disease, the bodily location affected, and the agency of disease, for instance, were based on the Type Of Occurrences Classification System (TOOCS) Third Edition of the Australian Safety and Compensation Council. Similarly, occupation classifications were based on the Australian and New Zealand Standard Classification of Occupations (ANZSCO) First Edition. Grouping intervals for age and employer size were

Table 3.1 Selected variables for data mining.

Variable	Measurement	Classification system followed
Size of employer	Micro (less than 5 employees); small (5–19 employees); medium (20–199 employees); large (above 200 employees)	Derived from ABS classifications
Age	Under 20; 20 to 29; 30 to 39; 40 to 49; 50 to 59; 60 & above	Derived from ABS classifications
Gender	Male; Female	Natural
Occupation	Air-conditioning and refrigeration mechanic; bricklayer/stonemason; building labourer; carpenter and joiner; concreter; crane/hoist/lift operator; earthmoving plant operator; electrician; electrical trade worker; fencer; forklift driver; glazier; industrial spray painter; insulation and home improvement installer; metal fitter and mechanist; painter; plasterer; plumber; railway track worker; roof tiler; sheet metal trade worker; structural steel construction worker; structural steel and welding trade worker; truck driver; wall and floor tiler.	ANZSCO classification
Nature of disease*	• Musculoskeletal and connective tissue diseases; • Digestive system diseases; • Skin disease; • Nerves system disease; • Respiratory system disease • Circulatory system disease (i.e. ischaemic heart disease, other heart disease; • Infectious disease; • Cancer.	TOOCS3.0 classification

Bodily location affected*	Head; Neck; Trunk; Upper limbs; Lower limbs; Multiple locations; Systemic locations; Psychological system (non-physical location); Unspecified location	TOOCS3.0 classification
Mechanism of disease*	Falls; Hitting objects with a part of the body; Being hit by moving objects; sound and pressure; body stressing; heat, electricity and other environmental factors; chemicals and other substances; biological factors; vehicle accident.	TOOCS3.0 classification
Agency of disease*	Machinery and fixed plant; mobile plant and transport; powered equipment, tools and appliances; non-powered equipment, hand tools and appliances; chemicals; materials and substances; environmental agencies; animal, human and biological agencies.	TOOCS3.0 classification
Disease severity	Fatality, permanent disability; temporary disability.	Safe Work Australia

Note
* The classifications have sub-classifications specific to the diseases. These will be examined in discussions on the disease type.

derived from the classifications used by the Australian Bureau of Statistics (ABS). Following that, 8250 records with zero compensation were removed. Further examinations were done to identify records with missing or invalid data for the filtered attributes and 10,302 such records were removed. This exercise resulted in a reduced dataset of 25,015 cases of occupational diseases for construction operatives.

Following that, descriptive statistics were computed to identify diseases that have a bigger health impact on workers. Cross-tabulation with chi-square analysis was performed for this purpose and Table 3.2 shows the results. It is evident

Table 3.2 Crosstab results

Summary

	Cases					
	Valid		Missing		Total	
	N	%	N	%	N	%
Disease * Severity	25,015	100.0	0	0	25,015	100.0

*Disease * Severity Crosstabulation*

Disease	Total count	Severity distribution (count)		
		Fatality	Permanent incapacity	Temporary incapacity
Musculoskeletal and connective tissue diseases	12,734	6	1072	11,656
Digestive system diseases	3220	n.p.*	74	3145
Skin and subcutaneous tissue diseases	1159	n.p.*	39	1120
Nervous system and sense organ diseases	7230	n.p.*	4935	2295
Respiratory system diseases	205	n.p.*	39	162
Circulatory system/cardio vascular diseases	172	17	18	137
Infectious and parasitic diseases	92	n.p.*	n.p.*	85
Neoplasms (cancer)	84	10	36	38
Other	119	n.p.*	16	102
Total (count)	25,015	42	6233	18,740

Chi-Square Tests

	Value	df	Asymptotic significance (2-sided)
Pearson chi-square	12,130.352[a]	18	0.000
Likelihood ratio	10,338.069	18	0.000
Linear-by-linear association	6045.611	1	0.000
No. of valid cases	25,015		

Notes
* Not publishable due to data sharing restrictions.
a Seven cells (23.3 per cent) have expected count less than 5. The minimum expected count is 0.04.

that the most common occupational disease suffered by construction operatives is musculoskeletal and connective tissue disorders, followed by nervous system and sense organ diseases. These two diseases account for almost 80 per cent of all the disease incidents. In addition, 96 per cent of permanent disabilities are suffered due to these two diseases. Nevertheless, fatalities occurred largely due to another two diseases. During the 13-year period that the dataset represents, almost two-thirds of the fatalities have been caused by cardiovascular diseases (circulatory system diseases) and neoplasms (cancer). The statistics in the crosstab help organise the four diseases in the following descending order of criticality: cardiovascular diseases, neoplasms (cancer), nervous system and sense organ diseases, and musculoskeletal and connective tissue disorders. While preventing every occupational disease is important, developing mechanisms to control these four disease categories is urgent in construction, given their criticality to the health of the operatives. Data analytics is therefore focused on these four occupational diseases in this chapter.

Techniques

Depending on the dataset size and the variable attributes, different techniques were applied to explore the data pertinent to the selected disease groups. The techniques applied include multiple correspondence analysis, classification trees and crosstab with cluster graphs. Details about the multiple correspondence analysis technique were discussed in Chapter 2. Crosstab with cluster graphs is a method for analysis via visualisation of association between two categorical variables. Classification trees is a data mining approach, which is further explained in the following paragraphs.

Classification trees is a widely used analytics technique in many fields, including medicine, computer science and psychology, as this produces outputs in an easily interpretable graphical format (Camdeviren *et al.* 2007). Moreover, it is a non-parametric procedure that is able to handle skewed datasets and does not require normally distributed data. Some of the most well-known machine-learning algorithms for building a classification tree include: chi-square automatic interaction detection (CHAID), classification and regression tree (CART); C5 and QUEST (Hajakbari and Minaei-Bidgoli 2014). CHAID is suitable for creating classification trees with categorical variables. It can uncover complex interactions among predictor variables that are most important in determining the outcome, based on if-then logic (Kim *et al.* 2010). This study applied CHAID since the dataset has been recorded with categorical variables. The CHAID algorithm operates using a series of merging, splitting and stopping steps as described below (Miller *et al.* 2014; Ramaswami and Bhaskaran 2010):

1   The merging step operates using each predictor variable where CHAID merges nonsignificant categories.
2   The splitting step occurs following the determination of all the possible merges of each predictor variable. CHAID recursively splits a population

(dataset) into separate and district segments. These segments, called nodes, are split in such a way that the variation of the response variable is minimised within the segment and maximised among the segments.

3   The splitting process is repeated on each node until one of the following stopping rules are met:

(a)   The tree reached the maximum depth level.
(b)   The size of a node is less than the user-specified minimum node size.
(c)   If the split of a node results in a child node whose node size is less than the specified minimum child node size, then the node will not be split.

The remainder of this chapter discusses the discoveries of data mining and analytics related to work-induced cardiovascular diseases, cancer, nervous system diseases and musculoskeletal disorders, using the aforementioned techniques.

Cardiovascular diseases among construction operatives

Cardiovascular diseases (CVD) refer to a variety of conditions that affect the heart and blood vessels (circulatory system), and are amongst the leading causes of death worldwide. These diseases are the results of a process called atherosclerosis, which is the build-up of fatty deposits (plaque) on the inside walls of arteries. Atherosclerosis can cause narrowing in, and possibly blockages of, arteries, which results in poor blood supply to vital parts of the body (Baker IDI 2011). Atherosclerosis can affect any artery in the body, including arteries in the heart, brain, arms, legs, pelvis and kidneys and, as a result, different diseases can develop based on which arteries are affected (University of Michigan Health System 2014). Poor blood flow to the heart is called coronary artery disease and can causes a heart attack (also referred to as ischaemic heart disease). Poor blood flow to the brain can cause a stroke (also known as cerebrovascular disease). Poor blood flow to the arms or legs is called peripheral artery disease (University of Ottawa Heart Institute 2011). The Australian Safety and Compensation Council (2008) also listed a few other diseases under the category of circulatory system diseases in TOOCS3.0 Classification, including: venous thromboembolism; hypertension (high blood pressure) and vibration white finger (Raynaud's disease).

Cardiovascular diseases are caused by a combination of genetic and lifestyle risk factors. The genetic risk factors include age, gender, heredity (family history of CVD) and ethnicity. Among the lifestyle risk factors are: smoking, overweight (especially around the waist), high blood cholesterol, high blood pressure, physical inactivity, excessive stress levels and depression (Baker IDI 2011; University of Ottawa Heart Institute 2011). These risk factors can be considered non-work related and applicable to the general population. There are numerous studies in the literature that investigate work-related risk factors for CVD and these factors are discussed under seven categories below.

- **Physical demands of work and fatigue:** physical exertion and lifting have been suggested as increasing the risk of CVD. While physical activity is required to reduce the risk of CVD, irregular heavy physical exertion can increase the risk (Hwang and Hong 2012). Steenland *et al.* (2000) claimed that temperature extremes and high physical demands increase the workload on the heart, resulting in acute coronary incidents. Collins (2009) stated that physical work demand combined with reduced job control (work organisation) leads to fatigue which, in turn, increases the risk of CVD, particularly hypertension.
- **Exposure to chemicals and lead:** occupational exposure to certain chemicals, such as carbon monoxide, carbon disulphide, methylene chloride and nitro-glycerine, has been conclusively related to CVD (Humblet *et al.* 2008). Likewise, occupational exposure to lead can promote atherosclerosis, hypertension and elevated blood pressure (Prokopowicz *et al.* 2017).
- **Exposure to noise and vibration:** an extended exposure to workplace noise and vibration has been shown to increase cholesterol concentration in blood, heart rate, arterial blood pressure, and thereby the development of hypertension and the risk of myocardial infarction and coronary heart diseases (Tomei *et al.* 2010; Lee *et al.* 2009; Davies *et al.* 2005).
- **Exposure to particulate matters:** exposure to ambient particulate air pollution is a recognised risk factor for cardiovascular disease. This is extendable to workplaces too. Fang *et al.* (2010) found an association between exposure to occupational particulate matters (fine particulates) such as silica, styrene, diesel exhaust, asphalt fumes, and metal and welding fumes, and adverse cardiovascular events. Their effects on the heart/circulatory system is not direct, but mediated through inflammation in the lungs, with the resulting release of fibrinogen into the blood stream leading to increased coagulability (Pope *et al.* 2004).
- **Passive smoking at workplace:** passive smoking at work occurs when exposed to indoor atmospheric pollution by tobacco smoke (Hwang and Hong 2012). Workers in bars, restaurants and nightclubs are particularly vulnerable to passive smoking. It has been proven that exposure to environmental tobacco smoke increases the heart rate and blood pressure, which leads to increased risk of CVD (Zhang *et al.* 2002).
- **Nightshift and overtime work:** Nowrouzi-Kia *et al.* (2018) found a significant statistical relationship between heart diseases and working an increased number of hours per week (as a result of working longer hours daily and working weekends) for a longer period (i.e. 10 years or more). Hartley *et al.* (2011) assessed that working nightshifts regularly increases the risk of CVD. The reasons for this were explained by other researchers that long working hours and nightshift work can increase blood pressure and thereby lead to increased risk of CVD, irrespective of other stressful conditions at work.
- **Occupational stress:** CVD risk is predicted by an array of risk factors, most prominently cholesterol, hypertension and smoking, and occupational stress

elevates the levels of these risk factors (Byrne and Espnes 2008). For example, elevated blood pressure has been related to occupational over-commitment, job strain and expression of workplace anger (Steptoe *et al.* 2004; Bongard and al'Absi 2005). Similarly, the prevalence of hypertension is related to organisational job constraints (Radi *et al.* 2005), excessive work hours (Yang *et al.* 2006), job insecurity and low occupational prestige (Levenstein *et al.* 2001), covert coping with unfair treatment at work (Theorell *et al.* 2000), perceived job barriers and job intensity (Greiner *et al.* 2004) and discriminatory treatment at work (Din-Dzietham *et al.* 2004). Other studies have established evidence associating occupational stress with harmful levels of blood lipids (cholesterol) (Kang *et al.* 2005). Finally, Metcalfe *et al.* (2003) and Ng and Jeffery (2003) noticed a particular emergence of association between occupational stress and a broad range of smoking. Byrne and Mazanov (2007) observed that occupational stress increases smoking frequency among established smokers.

Construction work is highly demanding, and workers are incessantly exposed to the occupational risk factors of CVD discussed above. Yet, there are few studies and little literature on CVD among construction operatives. Chung *et al.* (2018) examined the association between healthy behaviour and cardiovascular health among construction workers in Hong Kong and found that approximately two-thirds of the 626 workers sampled achieved only three out of seven 'ideal' cardiovascular health metrics. Similarly, Lunde *et al.* (2016) assessed that construction workers exhibit a higher level of cardiovascular load. This alarming evidence suggests that detailed studies are required to understand the prevalence and causes of CVD among construction operatives. Accordingly, the following section apprises evidence discovered from the workers' compensation data in Australia.

Multiple correspondence analysis was performed to investigate the relationships among CVD, disease severities, occupation and the age of the worker. Figure 3.1 shows the joint plot produced out of the analysis. A careful look at the distribution of variable scales suggests that the Y-axis represents worker characteristics and the X-axis represents diseases characteristics. The strength of association between the categories of the variables is estimated by the distance between the categories concerned. The following insights are drawn from the plot:

- As represented by the two quadrants on the upper part of the plot, the majority of CVD incidents in the construction industry resulted in temporary incapacity, and three diseases are heavily associated with these, namely: hypertension, venous disease and cerebrovascular disease (stroke).
- Temporary incapacities caused by stroke are more prevalent among industrial spray painters and glaziers. This could be due to their exposure to chemicals and particulate matters. Venous diseases affect operatives who work in sitting or squatting positions for long periods, and these trades

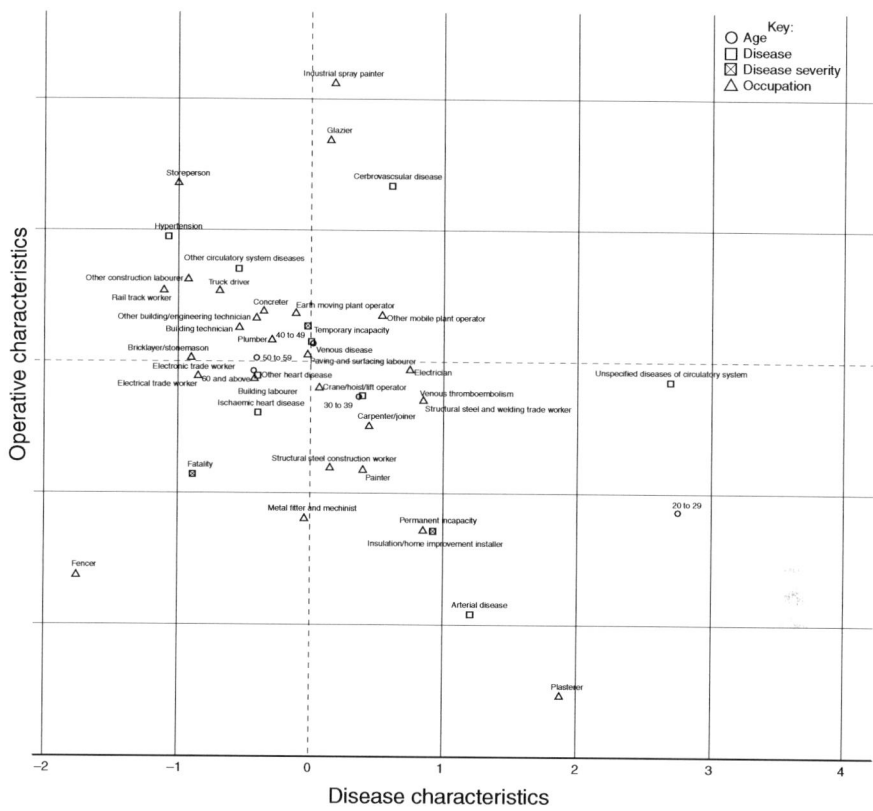

Figure 3.1 MCA results for CVDs.

include: plant operators, plumbers, and paving and surfacing labourers. Hypertension is more prevalent among truck drivers, construction labourers and rail track workers. This could be due to high physical demand and fatigue from the work performed.

- The lower right quadrant maps the associations related to permanent incapacities, and arterial diseases and venous thromboembolism appear to be the dominant causes. Moreover, painters, insulation/home improvement installers, structural steel and workers, carpenters/joiners and crane operators appear to be the more vulnerable work trades.

- The lower left quadrant displays the associations around CVD-related fatalities. Coronary diseases such as ischaemic heart diseases (heart attack) and other diseases that cause blockages in the arteries of the heart are closely related to fatalities among construction operatives. These diseases are more prevalent among building labourers, electrical trade workers, bricklayers/ stonemasons and metal fitters/machinists.

- In terms of relationships between the age of the operatives and CVD, workers aged 60 and above are more susceptible to more serious diseases, specifically heart attacks.

Neoplasms (cancer) among construction operatives

Cancer is a term generally used to group diseases of the body's cells. Normally cells in the human body grow and multiply in a controlled way. Cancer occurs if something causes a mistake in the cells' genetic blueprints, resulting in abnormal cell growth in an uncontrolled way. These new abnormal cells (referred to as neoplasms) can invade the surrounding tissues and spread to other parts and major organs of the body via the circulatory system. Most cancers start in a particular organ, which is called the primary site or primary tumour. When the abnormal cells do not spread beyond the primary site/immediate area, they are called benign and not dangerous. On the other hand, if these spread to different parts of the body, they are referred to as malignant – commonly known as cancer (Cancer Council 2019; Safe Work Australia 2014). There are many different types of cancers and usually they are named for the organ or cell type of the primary site; for example, lung cancer starts in the lung and stomach cancer starts in the stomach. Further, the type of cancer is associated with specific exposures; for example (Clapp *et al.* 2008, p. 1):

- Brain cancer from exposure to non-ionising radiation, particularly radiofrequency fields emitted by mobile phones;
- Breast cancer from exposure to pesticide dichloro-diphenyl-trichloroethane (DDT) prior to puberty;
- Leukaemia from exposure to 1,3-butadiene;
- Lung cancer from exposure to air pollution;
- Non-Hodgkin's lymphoma (NHL) from exposure to pesticide and solvents;
- Prostate cancer from exposure to pesticides, polyaromatic hydrocarbons (PAHs) and metal working fluids or mineral oils.

Occupational cancers occur due to exposure to carcinogens (cancer-causing agents) in the workplace, which damage the genetic blueprints of body cells. These agents are grouped under the following categories (Cancer Council 2015; Fernandez *et al.* 2012; Lacourt *et al.* 2015):

- Combustion products (e.g. engine exhaust, diesel, second-hand tobacco smoke);
- Industrial chemicals (e.g. benzene, vinyl chloride, formaldehyde, trichloroethylene, acid mist);
- Organic dusts (e.g. leather dust, wood dust, soil dust);
- Inorganic dusts (e.g. asbestos, silica dust, crystalline in the form of quartz or cristobalite, Portland cement, lime, gypsum);
- Metals (e.g. arsenic, cadmium, chromium, lead, nickel and their compounds);

- Radiation (e.g. artificial ultraviolet radiation, sun radiation, ionising radiation);
- Work behaviour pattern (i.e. shift work that involves circadian disruption).

The type of agent prevalent in the workplace, however, varies depending on the nature of work performed. Carcinogens specific to the construction industry and their associated cancer sites/types are listed in Table 3.3. It can be derived from the table that employment in the construction industry in general shows an elevated risk for lung cancer.

There have been studies that investigated the associations between cancer types and occupations in the construction industry, and Table 3.4 summarises the associations noted in the existing literature (Keller and Howe 1993; Pukkala *et al.* 1994; Knutsson *et al.* 2000). Nonetheless, occupational cancers are caused by past exposures and often there can be a period of many decades between exposure to a carcinogen and a subsequent cancer incident (Safe Work Australia 2016). Further, construction workers may be exposed to carcinogens that are used in their trades and/or from other trades in a shared work environment (Lacourt *et al.* 2015).

Table 3.3 Carcinogens pertinent to construction and associated cancers

Carcinogens	Cancer sites with strong evidence	Cancer sites with limited evidence
Asbestos	Mesothelioma, lung, larynx, ovary	Colon, rectum, stomach
Benzene	Leukaemia	Non-Hodgkin lymphoma
Chromium (VI) compounds	Lung	Nasal cavity and paranasal sinus
Engine exhaust, diesel	–	Lung, urinary bladder
Formaldehyde	Leukaemia, nasopharynx	Nasal cavity and paranasal sinus
Lead compounds	–	stomach
Nickel compounds	Lung, nasal cavity and paranasal sinus	–
Silica dust, crystalline (in the form of quartz or cristobalite)	Lung	–
Solar UV radiation	Skin	Eye
Second-hand tobacco smoke	Lung	Larynx, pharynx
Trichloroethylene	Renal cancer	Liver and biliary tract; non-Hodgkin lymphoma
Wood dust	Nasal cavity and paranasal sinus, nasopharynx	–

Source: modified from Safe Work Australia (2016).

Table 3.4 Cancers and occupations

Cancer type	Occupations associated
Kidney	Metal working; welding
Non-Hodgkin's lymphoma	General construction; metal working; painting; plumbing
Testis	General construction; painting; concreter
Nervous system	Electrician
Colon	Welding; concreter
Oral cavity	Electrician
Leukaemia	Welding
Liver	Cement work; carpenter
Stomach	Cement work; welding; concreter
Melanoma	General construction; plumbing
Pancreas	Plumbing
Oesophagus	General construction; painting; carpentry
Lung	General construction; metal working; welding; painting; bricklayer; concreter
Prostate	General construction; plumbing; metal working; welding; electrician; concreter
Bladder	General construction; electrician; painting
Rectum	Painting
Bone	Painting
Lip	Concreter

These two factors pose challenges in associating exposures to occupations. Nevertheless, Keller and Howe (1993) observed that construction workers experienced cancers earlier in life than other occupational groups.

Classification analytics of cancer data

Existing evidence or literature discusses the prevalence of cancer in the construction industry in general, but it does not relate the disease severity/outcome with the type of cancer suffered and other work and worker-related factors. Classification tree analytics was performed on past workers' compensation data to discover this pattern. Eighty-four records of occupational cancer were extracted from the workers' compensation claims between 2002 and 2014. Seven explanatory variables and one outcome variable were utilised for classification. The explanatory variables include: worker's age, gender, occupation, disease (cancer type), disease location in the body, disease mechanism and disease agent. The outcome variable is disease severity.

The classification tree was constructed using the CHAID algorithm with cross validation. CHAID partitioned the cancer disease dataset into statistically significant subgroups that were mutually exclusive. Accordingly, the classification tree shown in Figure 3.2 has three levels with a total of eight nodes, of which six are terminal nodes. Only three explanatory variables reached

Model Summary

Specifications	Growing Method	CHAID
	Dependent Variable	Disease severity
	Independent Variables	Age, Occupation, Disease, Disease mechanism, Bodily part affected
	Validation	Cross Validation
	Maximum Tree Depth	3
	Minimum Cases in Parent Node	10
	Minimum Cases in Child Node	5
Results	Independent Variables Included	Disease mechanism, Age, Occupation
	Number of Nodes	9
	Number of Terminal Nodes	6
	Depth	3

Classification performance

	Predicted			
Observed	Fatality	Permanent incapacity	Temporary incapacity	Percent Correct
Fatality	8	2	0	80.0%
Permanent incapacity	0	33	3	91.7%
Temporary incapacity	0	8	30	78.9%
Overall Percentage	9.5%	51.2%	39.3%	84.5%

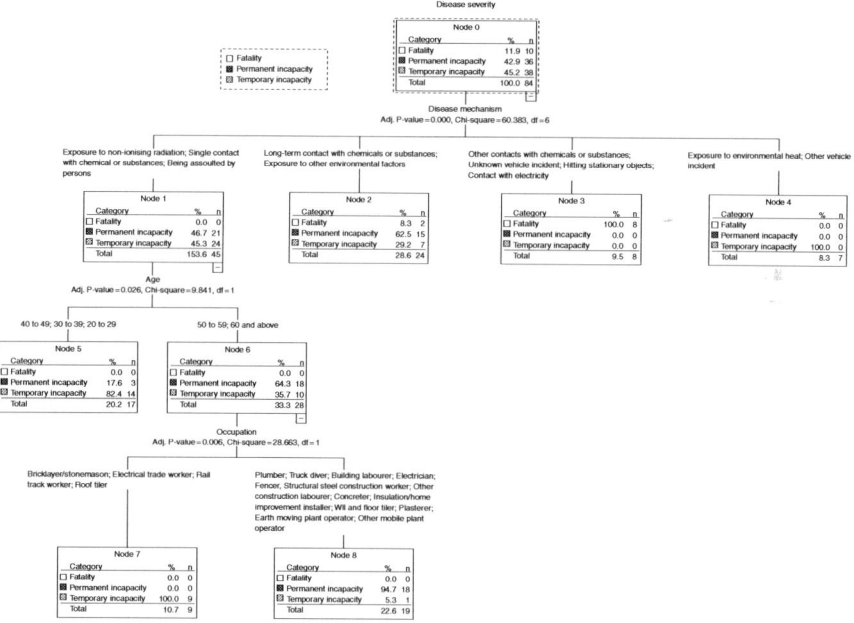

Figure 3.2 Classification analysis results for cancer.

statistical significance to be included in the model, namely disease mechanism, age and occupation. The variables such as disease and bodily part affected did not reach statistical significance for model construction. Moreover, the model had an overall classification accuracy of 84.5 per cent.

The tree starts with the root node (outcome variable) with information of the complete dataset and the spread of disease severities. The dataset is then split into four branches (nodes) at level one based on the statistically significant values of the explanatory variable, disease mechanism. Three out of the four branches are terminal nodes (Nodes 2, 3 and 4) with different disease severity outcomes. The nodes link the disease mechanisms with the predominant disease severity. It is evident from Node 3 that the type of carcinogens that caused fatalities is not very clear in the existing records. Node 2 reveals that long-term contact with chemicals and substances caused largely permanently incapacitating cancer incidents and a very low fraction of fatalities. Terminal Node 4 reveals that exposure to environmental heat only resulted in temporary incapacities. The type of disease suffered could be benign. However, Node 1 contains almost similar percentages of permanent and temporary incapacity cases and therefore these are further analysed in level 2.

At level 2, Node 1 is further split by the explanatory variable of age, resulting in two sub-branches with one terminal node (Node 5), which predominantly represents temporary incapacities. This may be interpreted that: workers aged between 20 and 50 mostly suffered benign cancer owing to exposure to non-ionising radiation and/or single contact with chemicals or substances. Workers aged 50 and above are further analysed in level 3 using occupation as the explanatory variable, which resulted in two terminal nodes, Nodes 7 and 8. The results can be interpreted as follows:

- Occupations such as bricklayer/stonemason, electrician, rail track worker and roof tiler are more likely to suffer temporary incapacities even if they are aged 50 and above. It could be because the effect of their exposure to non-ionising radiation and/or single contact with a chemical or substance is minimal, resulting in benign cancer.
- Workers aged 50 and above in occupations such as plumber, electrician, concreter, insulation installer, wall and floor tiler, plasterer, structural steel construction worker, plant operator and labourer are more likely to suffer permanent incapacities due to cancers caused by non-ionising radiation and/or single contact with a chemical or substance.

The CHAID machine learning algorithm did not establish the relationship between cancer types (disease) and disease mechanisms. However, a crosstab with chi-square statistic was performed to investigate the associations between the two variables. The associations were statistically significant (p-value =0.000) as shown in Table 3.5. It is evident in the table that construction workers in Australia are affected predominantly by two types of cancer, namely skin cancer and mesothelioma, which is a cancer that affect the mesothelial cells in the lungs, abdomen and many of the organs in the abdomen, heart and testicles (American Cancer Society

Table 3.5 Cancer type and exposure (%)

Disease mechanism	Cancer type							Total
	Mesothelioma	Melanoma of skin	Other skin cancer	Carcinoma in situ of skin	Other malignant neoplasms/carcinomas	Other benign neoplasms	Unspecified neoplasm	
Long-term contract with chemicals or substances	52.9				66.7			26.2
Single contact with chemical or substances	23.5							9.5
Other contacts with chemicals or substances	11.8							4.8
Exposure to other environmental factors		7.1	7.1					2.4
Contact with electricity	2.9							1.2
Exposure to non-ionising radiation		92.9	64.3	87.5			85.7	41.7
Exposure to environmental			28.6	12.5			14.	7.1
Being assaulted by persons					33.3			2.4
Unknown vehicle incident	5.9							2.4
Other vehicle incident						100.0		1.2
Hitting stationary objects	2.9							1.2
Total	100.0	100.0	100.0	100.0	100.0	100.0	100.0	100.0

2018). In summary, two disease mechanisms, such as long-term contact with chemicals or substances and exposure to non-ionising radiation, appear to be more prevalent in the Australian construction industry and these are responsible for causing mesothelioma and skin cancer, respectively.

Nervous system and sensor organ diseases among construction operatives

A subset containing 7230 records of work-induced nervous system and sensor organ diseases was extracted from workers' compensation claims between 2002 and 2014. The following diseases were used by the TOOCS to record work-induced nervous system and sensor organ diseases in Australia:

- Diseases of the brain, spinal cord and peripheral nervous system
- Diseases of nerve roots, plexuses and single nerves
- Carpal tunnel syndrome
- Diseases of the conjunctiva and cornea
- Other diseases of the eye
- Deafness
- Other diseases of the ear
- Other diseases of the nervous system and sense organs
- Unspecified diseases of the nervous system and sense organs

A classification tree was built to discover the relationships between the disease severity and explanatory factors such as worker's age, gender, occupation, disease, disease location in the body, disease mechanism and disease agent. The classification tree was constructed using the CHAID algorithm with cross validation and is shown in Figure 3.3. The classification tree has three levels with a

Model Summary

Specifications	Growing Method	CHAID	
	Dependent Variable	Disease severity	
	Independent Variables	Age, Occupation, Disease, Bodily part affected, Disease mechanism, Disease agency	
	Validation	Cross Validation	
	Maximum Tree Depth		3
	Minimum Cases in Parent Node		100
	Minimum Cases in Child Node		50
Results	Independent Variables Included	Disease, Disease agency, Age, Disease mechanism	
	Number of Nodes		15
	Number of Terminal Nodes		11
	Depth		3

Classification accuracy

		Predicted	
Observed	Permanent incapacity	Temporary incapacity	Percent Correct
Permanent incapacity	4654	281	94.3%
Temporary incapacity	416	1879	81.9%
Overall Percentage	70.1%	29.9%	90.4%

Growing Method: CHAID
Dependent Variable: Disease severity

Figure 3.3 Classification analysis results for nervous system and sensor organ diseases.

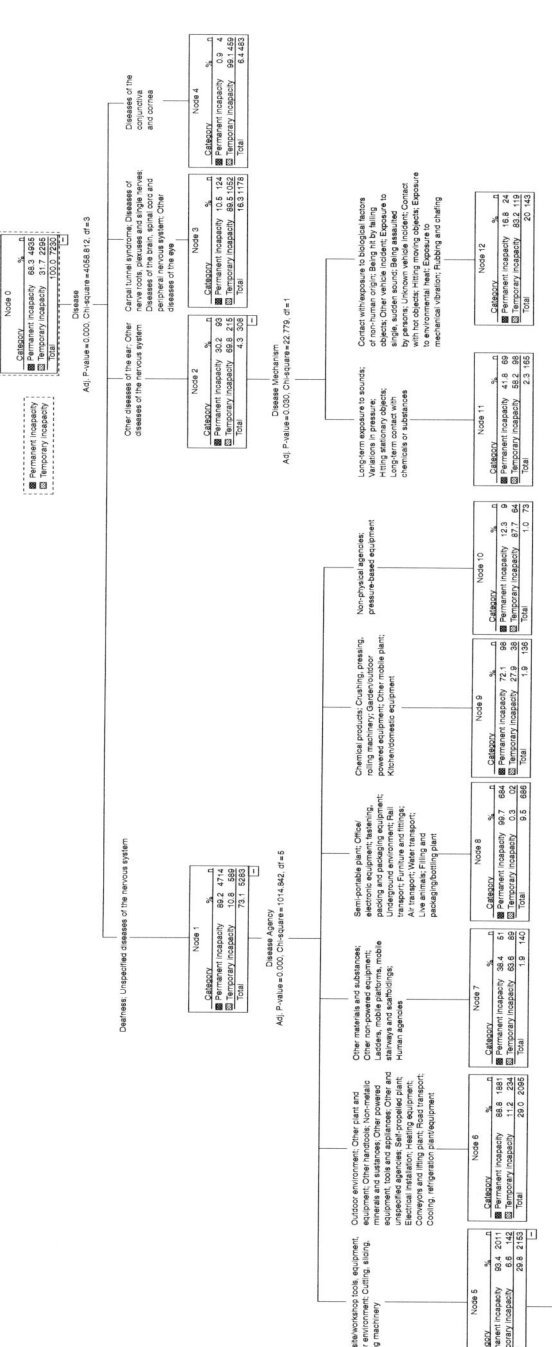

Figure 3.3 Continued

total of 15 nodes of which 11 are terminal nodes. Only four explanatory variables reached statistical significance to be included in the model, namely: disease, disease mechanism, disease agency and age. The variables such as bodily part affected and occupation did not reach statistical significance for model building. Moreover, the model had an overall classification accuracy of 90.4 per cent.

The classification tree starts with the root node (outcome variable) with information about the complete dataset and the spread of disease severities. There were no fatalities recorded but permanent (68 per cent) and temporary (32 per cent) incapacities. The dataset is then split into four branches (nodes) at level 1, based on the statistically significant values of the explanatory variable, disease. The nodes link the disease with the predominant disease severity. It is evident from the branch that deafness is the predominant cause of permanent incapacities (89 per cent) under the category of nervous system and sensor organ diseases (see Node 1). Node 1 is further split at level 2 by disease agency, resulting in six nodes, of which four represent permanent incapacities. A close observation of the agencies that are linked to permanent incapacities reveals that plant and machinery, tools and work environment are largely responsible for causing permanent deafness/hearing loss. Essentially, it is the noise that

Table 3.6 Categories of construction tools and equipment

Tool and equipment group	Included tools and equipment
Drills, wrenches and power guns	Handheld drills; masonry drills; impact wrenches; nail guns
Manual hand tools	Hammers; sledge hammers; chisel; hand saws; scraper/sander; trowels
Construction vehicles and trucks	Excavator; front end loaders; bulldozers; pipe laying vehicles and trucks
Spraying, vacuuming or blowing systems	Handheld air blower; spraying system; vacuum system
Planers, sanders and grinders	Sanders; grinders; power scraping trowels; planers
Saws	All saws including circular, concrete, metre, band, radial/drop, reciprocating, table and bench
Large machinery and power hammers	Compactor; rammers; rollers; tampers; chipping hammers; jack hammers
Concrete tools, welding and other	Concrete pumps; concrete trucks; concrete vibrators; welding equipment; other construction equipment that makes noise above 80 dB(A)
Background and passive noise	Equipment noise from others on site and onsite plant like pumps and generators

Source: modified from Lewkowski *et al.* (2017).

emanates from the operations on construction sites that causes hearing loss. Noise is a highly prevalent occupational hazard, particularly in the construction industry where noise levels frequently exceed standards (Seixas *et al.* 2005) and construction workers are at higher risk of noise exposure (Masterson *et al.* 2013). Long-term exposure to daily noise levels greater than 80 dBA, which European Directive 2003/10/EC defines as lower action level, is the principal cause for noise-induced hearing loss among construction operatives (Leensen *et al.* 2011). The last level in the classification tree (see Figure 3.3) analyses the spread of age groups and it is evident that two-thirds of the incidents occurred to workers under 60, providing evidence against the notion that age is a significant contributor to hearing loss among construction workers. Nonetheless, the age groups range from 20 to 50, making it difficult to identify the most affected cohort. It is, however, clear that workers aged 60 and above account for one-third of the noise-induced hearing loss incidents. Hearing loss is gradual, and it increases with the length of time on the job (CPWR 2018). Higher exposure levels for longer exposure durations cause greater hearing impairment (Dobie 2007). A high representation of workers aged 60 and above can possibly be related to the length of their employment in the construction industry.

Noise on construction sites is complex and dynamic. The noise exposure of construction workers varies greatly in a single day, with the varieties of activities performed, and tools and equipment used (Leensen *et al.* 2011). Lewkowski *et al.* (2017) categorised tools and equipment used on construction sites into eight groups, as shown in Table 3.6. They further investigated identifying tool groups that create noise levels above 80 dbA and that are widely used on sites. The results revealed four such groups, namely: planers, sanders and grinders; large machinery and power hammers; saws; and hand tools. In a single day, a worker may use several of these tools and items of equipment, just one continuously, use none or be exposed to noise from the equipment/tools used by other workers on site. Moreover, workers' daily tasks, environment and shift length may vary daily. All these, combined, make tracking and controlling noise exposure difficult at site or for individual workers. As such, hearing protection devices are often used as a viable means for reducing noise exposure on construction sites (Lewkowski *et al.* 2017). Since noise can affect any worker on the vicinity of a noisy operation, all workers in that zone should be provided and directed to wear hearing protection. For example, if other trade workers are working in a zone where concrete hacking is taking place, not only the person performing the task but also everyone in that work zone should be provided with and directed to wear hearing protection.

Hearing protection devices, such as ear plugs and ear muffs, can be effective for preventing hearing loss only if used consistently (Groenewold *et al.* 2014). However, several studies have shown that in the construction industry they are not used consistently. Lewkowski *et al.* (2017) quoted one US study which found that workers use hearing protection only a quarter of the time they were exposed to noise. In another study, also from US, the majority of workers exposed to hazardous noise did not wear hearing protection. In Australia, one study found that only 33 per cent of workers who were exposed to a full shift limit of 85 dB did use hearing protection at all on the day of the study. Discomfort, hindrance to communication, highly variable noise levels, which are common on construction sites, and awareness/education level of workers cause irregular or non-use of hearing protection by construction workers (Neitzel and Seixas 2005).

In summary, it is evident that noise-induced hearing loss is the predominant work-induced nervous system and sensor organ disease among construction workers. There is strong evidence that exposure to noise levels exceeding 80 dB(A) for a longer period results in hearing loss, which impacts on workers' overall quality of life and their safety on site. Continuous use of hearing protection devices such as ear plugs and ear muffs by workers is a practical solution to reduce the risk. Workers should be constantly trained and made aware of the criticality of using them for their health and safety, despite any discomfort or inconvenience that they may feel in wearing them.

Musculoskeletal disorders

Work-related musculoskeletal disorders (MSDs) are inflammatory and degenerative conditions that affect the muscles, tendons joints, nerves and supporting blood vessels that occur owing to work-related activities. Skeletal disorders include fractures, fracture of vertebral column with or without spinal cord lesion, dislocation, arthropathies disorder of joints, dorsopathies (disorder of the spinal vertebrae and the intervertebral discs), osteopathies (disorders of the bones), chondropathies (disorders of the cartilage) and acquired musculoskeletal deformities. Muscular disorders include strains and sprains of joints and adjacent muscles, disorders of muscle, tendons and other soft tissues and hernia (Australian Safety and Compensation Council 2006; Wang *et al.* 2016).

Previous research reported that in the construction industry, musculoskeletal disorders account for about one-third of all injuries and illnesses (Choi *et al.* 2016) and for more than 40 per cent of days lost from work (Sobeih *et al.* 2006). Musculoskeletal disorders have also been found to be a determinant of early retirement or disability in studies by LeMasters *et al.* (2006) and Oude Hengel *et al.* (2012) as quoted in Boschman *et al.* (2012). These statistics are largely US based and some are a decade old. Analyses were performed to investigate the prevalence of MSDs in the Australian construction industry. A dataset containing 12,725 records of work-induced musculoskeletal disorders among construction operatives in Australia was extracted from the workers' compensation claims between 2002 and 2014. Descriptive analysis of the data was first performed to examine the disease severity distribution. As shown in Figure 3.4, more than 90 per cent of the cases were recorded for temporary incapacities. Since the distribution does not exhibit a significant variability of disease severities, the application of classification tree analysis or multiple correspondent analysis was deemed not viable. Instead, the analysis proceeded with chi-square tests with cluster graphs to investigate

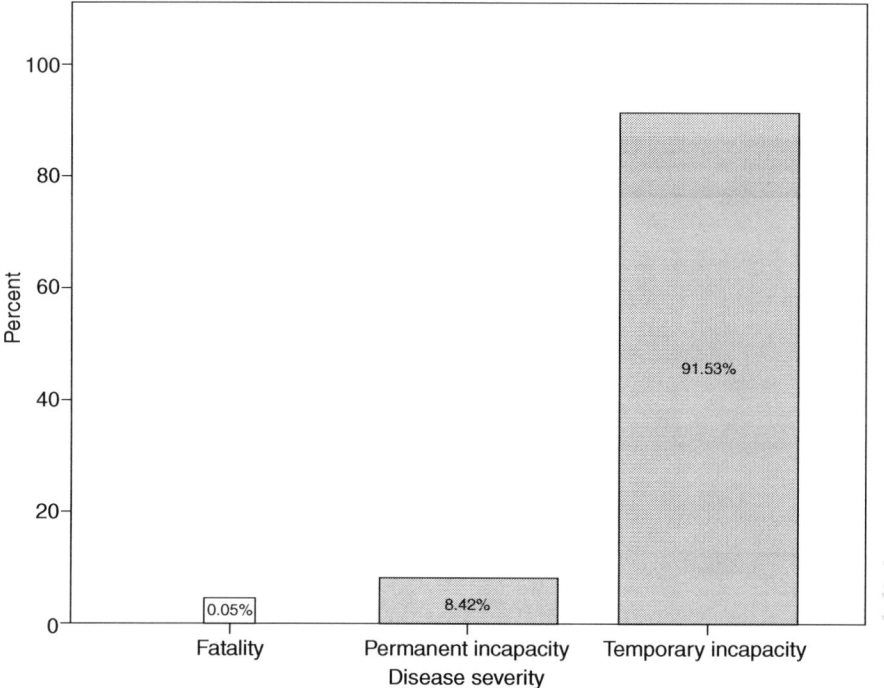

Figure 3.4 Distribution of musculoskeletal disorder severities.

which scales of the variables are heavily associated with work-induced musculoskeletal disease severities. The results revealed statistically significant variabilities only across the scales of five variables, namely: occupation, age, disease type, affected body part and disease mechanism. The findings are discussed below.

Occupation

Figure 3.5 shows the musculoskeletal disorder severities among different occupation groups in the construction industry. It is evident from the chi-square results (*p*-value <0.05) that the differences are statistically significant. Among the 40 occupation groups in the construction industry, the top ten sufferers of MSDs are identified and listed in descending order, as follows: (1) carpenters; (2) plumbers; (3) electricians; (4) structural steel construction workers; (5) concreters; (6) building labourers; (7) truck drivers; (8) plasterers; (9) painters; and (10) earth moving plant operators. Permanent incapacities due to MSDs are also the highest among carpenters and building labourers. Previous studies suggested that construction labourers and carpenters account for about 31 per cent, but in terms of incident rates (*x*/10,000), helpers (100.9), heating A/C

Chi-Square test results			
	Value	df	Asymptotic Significance (2-sided)
Pearson Chi-Square	362.521[a]	78	0.000
Likelihood Ratio	181.404	78	0.000
Linear-by-Linear Association	16.491	1	0.000
No. of Valid Cases	12725		
a. 49 cells (40.8%) have expected count less than 5. The minimum expected count is 0.01.			

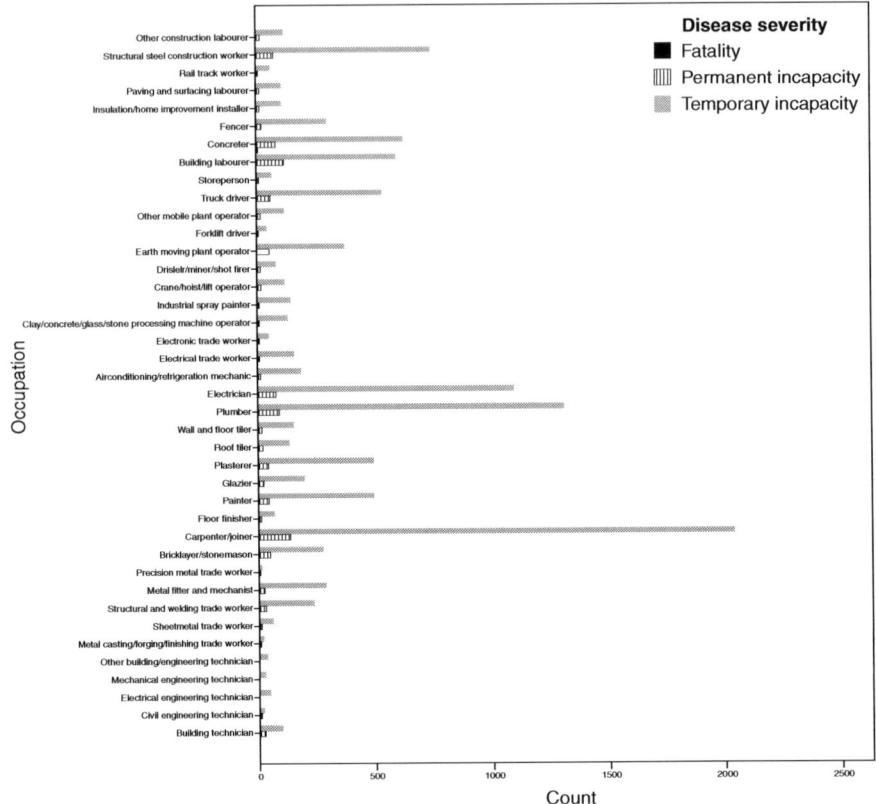

Figure 3.5 Occupation and musculoskeletal disorders.

mechanics (98.9) and cement masons (86.3) topped the list (Wang *et al.* 2016). However, the findings from this study produce a different list of vulnerable occupation groups.

Age

Figure 3.6 reveals the musculoskeletal disorder severities among different age groups of construction operatives. It is evident from the chi-square results (p-value <0.05) that statistically significant differences exist across different age

Chi-Square Tests			
	Value	df	Asymptotic Significance (2-sided)
Pearson Chi-Square	27.066[a]	10	0.003
Likelihood Ratio	29.453	10	0.001
Linear-by-Linear Association	17.951	1	0.000
No. of Valid Cases	12720		
a. 6 cells (33.3%) have expected count less than 5. The minimum expected count is 0.14.			

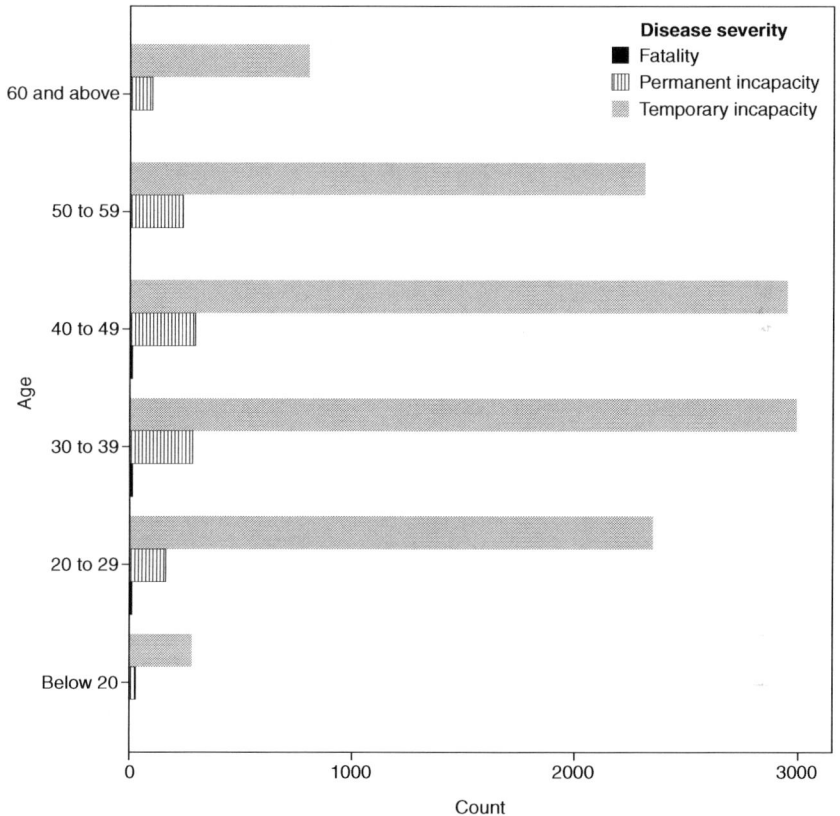

Figure 3.6 Age and musculoskeletal disorders.

groups in the severities suffered. The cluster graph shows a bell-curve pattern, meaning workers aged below 20 and above 60 are less represented in MSD records. Workers aged 30 to 49 are heavily represented in MSD, together accounting for almost 50 per cent of all MSD incidents, followed by workers aged 20 to 29 and 50 to 59, who are almost equally represented (i.e. almost 20 per cent for each of this age group, totalling around 40 per cent). Quite surprisingly, workers aged 60 and above are less represented in MSDs although their old age may be a contributory factor for this type of diseases. These findings are slightly different from previous studies by others in some respects but largely concur. Occupational Disease Indicators (Safe Work Australia 2014) reported that there are differences among the different age groups in terms of MSD prevalence and severities. According to them, the age groups 25 to 34 and 35 to 44 top the list, together accounting for 60 per cent of the total, which broadly reflects the age structure of the people working in the industry. They further suggested that, in terms of rates there was a difference between the 25 to 34 (35.3), 35 to 44 (43.9), 45 to 54 (47.1) and 55 to 64 (37.8) age groups while younger and older workers had lower rates.

Disease types and affected body parts

The Type Of Occurrences Classification System (TOOCS) Third Edition of the Australian Safety and Compensation Council, groups work-related MSDs under five broad categories, namely: joint and cartilage diseases; spinal disc diseases; diseases of synovium; diseases of muscle, tendon and related tissues; and other MSDs. Figure 3.7 illustrates the categories of MSDs that are common among construction operatives. The two common MSDs are: spinal disc diseases and the diseases of muscle, tendon and related tissues. These two together are responsible for over 70 per cent of the temporary and permanent incapacities due to MSDs among construction operatives.

Figure 3.8 coincides with the findings above, whereby it shows the predominantly affected body parts due to MSDs, which are: back, shoulder, knee, elbow, neck, wrist and hand. The variations in MSD occurrences and severities among the different body parts affected are statistically significant as evidenced by the chi-square results (p-value <0.05). Choi *et al.* (2016) previously reported that the predominantly affected body parts due to MSDs across all occupation groups in construction are back, shoulder and neck. According to this study, around 60 per cent of temporary incapacities and more than 80 per cent of the permanent incapacities due to MSDs among construction operatives are caused by damage to these three body parts. Permanent and temporary incapacities due to damage to body parts such as knees, and the combination of elbows, wrists and hands are almost equally represented in MSD severity statistics. They each account for about 12 per cent of temporary incapacity and 10 per cent of permanent incapacity incidents. In other words, almost one-quarter of temporary incapacities due to MSDs are caused by damage to knees, elbows, wrists and hands. Similarly, almost one-fifth of permanent incapacities due to MSDs are caused by damage to these body parts.

Disease mechanisms

Figure 3.9 identifies the task factors or disease mechanisms that are strongly associated with MSDs among construction operatives. Muscular stress due to lifting, carrying, handling loads or without involving loads, and repetitive

Chi-Square tests results			
	Value	df	Asymptotic Significance (2-sided)
Pearson Chi-Square	16.904ᵃ	8	0.031
Likelihood Ratio	18.169	8	0.020
Linear-by-Linear Association	0.084	1	0.772
No. of Valid Cases	12725		
a. 5 cells (33.3%) have expected count less than 5. The minimum expected count is 0.20.			

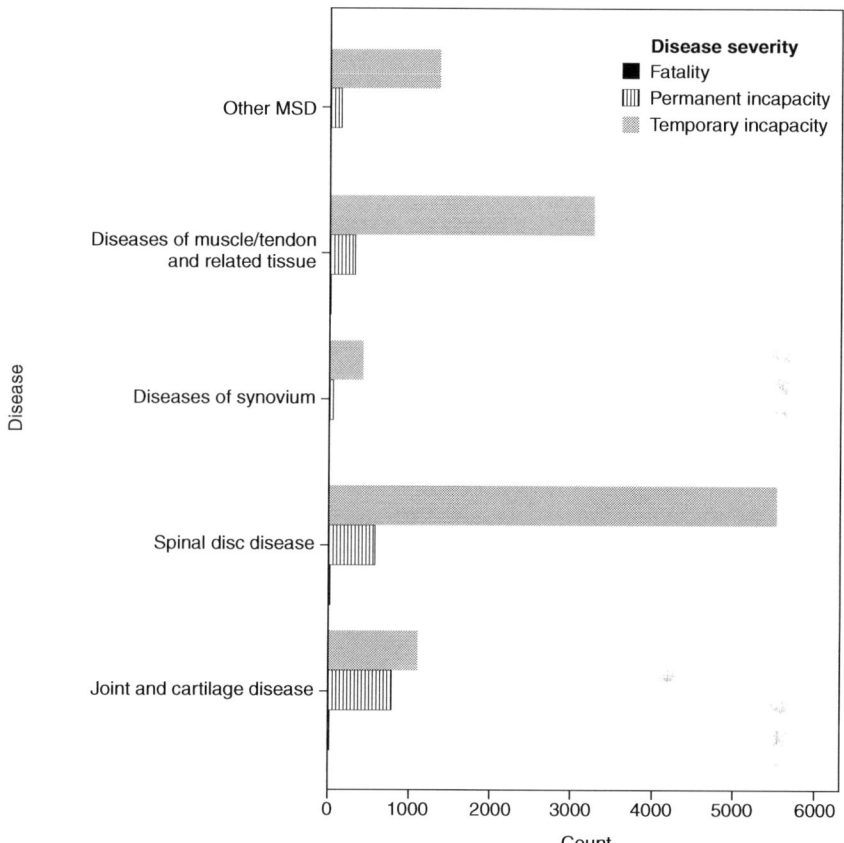

Figure 3.7 Musculoskeletal disorder types and severities.

movements are associated with about two-thirds of all temporary incapacities and more than 80 per cent of permanent incapacities caused by MSDs among construction operatives. The remaining portion of the MSDs are caused by: falls; being struck by objects or vehicles; stepping, kneeling or sitting on construction tasks; and exposure to mechanical vibration.

Musculoskeletal disorders among workers are traditionally associated with causes such as overexertion, repetitive motions and vibrations (Wang *et al.* 2016). This study also found similar associations with physical factors at work. However, Sobeih

Chi-Square test results			
	Value	df	Asymptotic Significance (2-sided)
Pearson Chi-Square	173.076[a]	54	0.000
Likelihood Ratio	162.214	54	0.000
Linear-by-Linear Association	8.914	1	0.003
No. of Valid Cases	12725		
a. 43 cells (51.2%) have expected count less than 5. The minimum expected count is 0.00.			

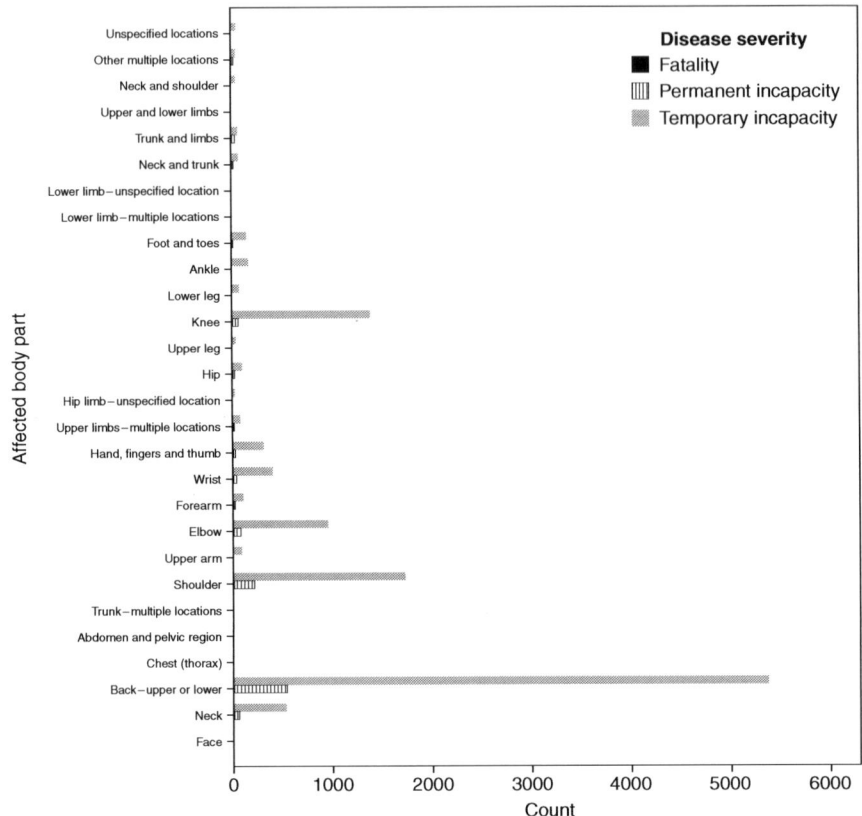

Figure 3.8 Affected body parts and musculoskeletal disorders.

et al. (2006) summarising some 20 research projects concluded that there is a positive association between psycho-social factors and musculoskeletal disorders too. They found strong evidence for low job satisfaction followed by high perceived job stress and musculoskeletal disorders. There was also a strong relationship between worry, distress and stress reactions, and musculoskeletal disorders, while low social support and low stimulus from work were not statistically significantly associated in a univariate analysis. This suggests that efforts aimed to curtail MSD among construction operatives should not only focus on physical factors but also psycho-social factors at work.

Chi-Square test results			
	Value	df	Asymptotic Significance (2-sided)
Pearson Chi-Square	127.298[a]	54	0.000
Likelihood Ratio	144.125	54	0.000
Linear-by-Linear Association	18.026	1	0.000
No. of Valid Cases	12725		
a. 48 cells (57.1%) have expected count less than 5. The minimum expected count is 0.00.			

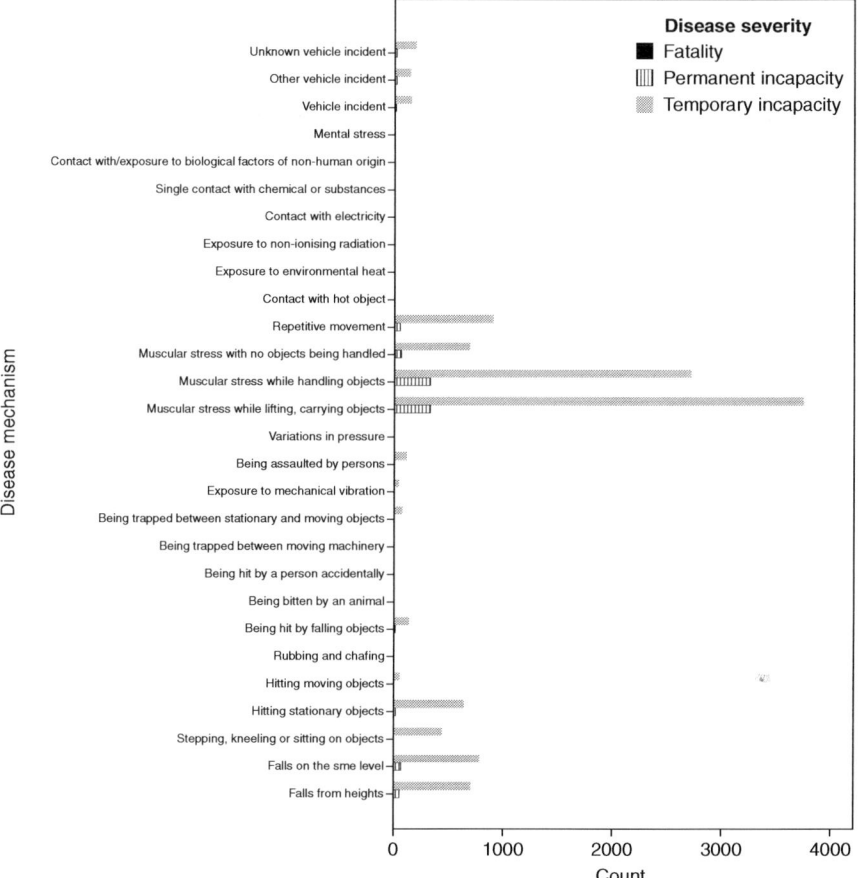

Figure 3.9 Incident mechanisms and musculoskeletal disorders.

MSDs and occupational risk factors

The account above examined MSDs among construction operatives using cluster graphs, taking one variable at a time. Table 3.7 integrates the associations between occupation types, nature of work, work-related MSD risk factors and body parts affected. The information presented in the table is drawn from this study as well as existing literature (Choi *et al.* 2016, p. 30; Choupani *et al.* 2015;

Table 3.7 Construction work trades and MSD hazards

Occupation	Nature of work	MSD hazards	Body parts affected
Carpenters	• Overhead work • Ground/floor-level work • Hand-intensive work • Manual material handling	• Forceful exertions • Awkward body postures • Bending/twisting the body • Repetitive motions • Hot/cold temperatures • Kneeling, crouching, stooping or crawling • Exposure to whole body vibration	• Back • Neck • Shoulders • Wrists/hands/fingers • Knees
Plumbers, pipe fitters and steam fitters	• Overhead work • Ground/floor-level work • Hand-intensive work • Manual material handling	• Force • Awkward body postures • Work in static positions • Pressure/pinch points • Hot/cold temperatures	• Back • Neck and shoulder • Elbows • Wrists/hands • Knees
Electricians	• Overhead work • Ground/floor-level work • Hand-intensive work • Manual material handling	• Force (pushing/pulling wires, bending conduits) • Awkward body postures • Bending/twisting the body • Repetitive motions • Hot/cold temperatures • Kneeling, crouching, stooping or crawling • Exposure to whole body vibration	• Back • Neck and shoulder • Wrists/hands/fingers
Structural steel construction workers	• Elevated level work • Hand-intensive work • Manual material handling	• Force • Awkward body postures • Work in static positions • Bending/twisting the body • Repetitive motions • Hot/cold temperatures • Kneeling, crouching, stooping or crawling • Exposure to whole body vibration	• Back • Neck and shoulders • Arms/Elbows • Wrists/hands • Knees

Occupation	Tasks	Risk factors	Body parts affected
Concreters	• Shovelling and screeding concrete • Use of hand-held concrete vibrators and other tools • Elevated/sloped level work • Ground/floor-level work • Manual handling • Prolonged standing position • Difficult working platform	• Awkward body postures • Bending/twisting the body • Repetitive motions • Exposure to whole body vibration • Exerting force	• Shoulder • Back • Arms/elbows • Hands/wrists • Knees • Legs • Feet
Building labourers	• Overhead work • Ground/floor-level work • Manual lifting and carrying heavy objects • Pushing, pulling, tugging and sliding • Work in hot, cold or wet conditions • Work in confined spaces • Using hand-held tools	• Awkward body postures • Bending/twisting the body • Repetitive motions • Kneeling, crouching, stooping or crawling • Exposure to whole body vibration	• Back • Neck • Shoulders • Elbows • Hands/wrists • Knee • Legs • Feet
Truck drivers	• Operating and manoeuvring the truck on construction sites • Loading, moving and unloading heavy loads • Constant use of pedals (i.e. brake, clutch and accelerator) • Constant use of hands and wrists to operate the truck • Field repairs and maintenance of trucks	• Prolonged seated position • Uncomfortable seats • Exposure to whole body vibration • Repetitive motions • Forceful exertions • Bending and twisting the body and neck	• Neck • Shoulder • Lower back • Knee • Feet • Hand/wrist
Plasterers	• Working above shoulder height • Outstretching • Work below knee level • Prolonged standing • Applying hand force	• Awkward postures • Repetitive motions • Bending/twisting the body	• Shoulders • Neck • Back • Elbows • Wrist/hands

continued

Table 3.7 Continued

Occupation	Nature of work	MSD hazards	Body parts affected
Painters	• Working above shoulder height • Work below knee level • Prolonged standing • Holding the hand brush or roller brush above the arm level for extended periods • Reaching out • Use of hand-held tools	• Awkward body postures • Bending/twisting the body • Repetitive motions • Kneeling, crouching, stooping or crawling	• Shoulders • Neck • Back • Elbows • Wrists/hands
Earthmoving plant operators	• Operating and manoeuvring the plant on construction sites • Moving and lifting loads • Prolonged work head/below floor level	• Prolonged seated posture • Exposure to whole body vibration • Awkward body postures • Repetitive motions • Bending and twisting the body and neck	• Arms • Shoulders • Neck • Lower back

CPWR 2017; Lin *et al.* 2015; Mozafari *et al.* 2015; Kittusamy and Buchholz 2004). The table discusses hazards encountered by the top-ten vulnerable occupations that were identified in this study.

Conclusion

Unlike for work-related injuries and fatalities, receiving fair workers' compensation for occupational diseases in the construction industry is challenging. This is predominantly due to the difficulties in proving that work-related exposure to/causes of diseases occurred in a particular project or employment, as there may be a time lag between the disease onset and the end of the project or employment. Moreover, a disease can be caused by work-related and/or non-work-related exposure. Maintaining an up-to-date, construction industry-specific registry of occupational diseases and affected occupation groups would, to some degree, ease the difficulties inherent in proving work-related causes for obtaining compensation. This chapter, through data analytics, established associations between four major groups of diseases and construction occupations, which are: cardio-vascular diseases, cancer, nervous system and sensor organ diseases, and musculoskeletal disorders. The findings may be applied to develop preventive measures of these diseases among construction operatives, whether through engineering or administrative strategies.

The challenge in relating a disease to a previous occupational exposure or employment and obtaining workers' compensation for it, however, remains. An additional mechanism is required to address this. That is, if a disease occurs within a certain timeframe of exposure to relevant hazards in a project, then the diseases may be considered to have arisen out of the employment or project. The time lag, also known as the latency period, may vary according to the nature and type of the disease. Table 3.8 suggests allowable time lags/latency

Table 3.8 Latency periods

Disease	Latency period	Source
Ischemic heart disease	10 to 30 years	Simonetto *et al.* (2014)
Skin cancer	2 years	Gawkrodger (2004)
Mesothelioma (lung or abdominal cancer)	10 to 50 years	Frost 2013; Reid *et al.* (2014)
Noise-induced hearing loss	Hearing loss is gradual and 5 years or more of exposure is generally required for significant hearing loss to occur. Hearing loss does not progress after noise exposure is discontinued.	Mathur (2018)
Musculoskeletal disorders	12 years	Nicoletti and Battevi (2008)

periods for critical occupational diseases predominant among construction workers as found in this chapter. The suggested latency periods were drawn from scientific literature.

In conclusion, in order for construction operatives to be compensated fairly for occupational diseases, the workers' compensation system should consider both work-specific exposure and the possible latency period between the exposure and the onset of disease symptoms, which may extend well beyond the actual duration of a construction project.

References

American Cancer Council (2018). What is malign mesothelioma? www.cancer.org/cancer/malignant-mesothelioma/about/malignant-mesothelioma.html (accessed 20 November 2018).

Australian Safety and Compensation Council (2006). Work-related musculoskeletal disease in Australia. www.safeworkaustralia.gov.au/system/files/documents/1702/workrelatedmusculoskeltaldisorders_2006australia_2006_archivepdf.pdf (accessed 10 October 2018).

Baker IDI (2011). Cardiovascular disease. www.baker.edu.au/Assets/Files/Cardiovascular Disease.pdf (accessed 5 July 2018).

Boden, L.I. (1986). Problems in occupational disease compensation. In J. Chelius (Ed.), *Current Issues in Workers' Compensation*. Kalamazoo, MI: W.E. Upjohn Institute for Employment Research, pp. 313–326. https://doi.org/10.17848/9780880995443.ch13.

Bongard, S. and al'Absi, M. (2005). Domain-specific anger expression and blood pressure in an occupational setting. *Journal of Psychosomatic Research*, 58(1): 43–49.

Boschman, J.S., van der Molen, H.F., Sluiter, J.K. and Frings-Dresen, M.H.W. (2012). Musculoskeletal disorders among construction workers: a one-year follow-up study, *BMC Musculoskeletal Disorders*, 13(196): 1–9.

Byrne, D.G. and Espnes, G.A. (2008). Occupational stress and cardiovascular disease. *Stress and Health*, 24(3): 231–238.

Byrne, D.G, and Mazanov, J. (2007). Personality, stress and the determination of smoking behaviour in adolescents. In G. Boyle, G. Matthews and D. Saklofske (Eds), *Handbook of Personality Theory and Testing*. London: Sage.

Camdeviren, H.A., Yazici, A.C., Akkus, Z., Bugdayci, R. and Sungur, M.A. (2007). Comparison of logistic regression model and classification tree: an application to postpartum depression data. *Expert Systems with Applications*, 32(4): 987–994.

Cancer Council (2019). What is cancer? www.cancer.org.au/about-cancer/what-is-cancer/ (accessed 27 April 2019).

Cancer Council Western Australia (2015). Occupational exposures to carcinogens in Australia. www.cancer.org.au/content/pdf/News/MediaReleases/2015/web%20%20Occupational%20report.pdf (accessed 20 March 2018).

Centre for Construction Research and Training (CPWR) (2017). *Chart Book*, 6th edn. *Hazards and Exposures – Exposure Risks for Work-Related Musculoskeletal Disorders and Other Illnesses in Construction*. www.cpwr.com/chart-book-6th-edition-hazards-and-exposures-exposure-risks-work-related-musculoskeletal-disorders (accessed 10 October 2018).

Centre for Construction Research and Training (CPWR) (2018). Topics in construction safety and health: noise and hearing loss. www.cpwr.com/sites/default/files/publications/Noise-and-Hearing-Loss-annotated-bibliography.pdf (accessed 16 March 2018).

Choi, S.D., Yuan, L. and Borchardt, J.G. (2016). Musculoskeletal disorders in construction: practical solutions from the literature. *Professional Safety*, 61(1): 26–32.

Choupani, A., Hosseini, S.Y., Sadeghi, M. and Namdari, M. (2015). Musculoskeletal disorders among plastering workers. *BEMS Reports*, 1(1): 16–19.

Chung, J.W., Wong, B.Y., Yan, V.C., Chung, L.M., So, H.C. and Chan, A. (2018). Cardiovascular health of construction workers in Hong Kong: a cross-sectional study. *International Journal of Environmental Research and Public Health*, 15(6): 1251–1272.

Clapp, R.W., Jacobs, M.M. and Loechler, E.L. (2008). Environmental and occupational causes of cancer: new evidence 2005–2007. *Reviews on Environmental Health*, 23(1): 1–37.

Collins, S. (2009). Occupational factors, fatigue and cardiovascular disease. *Cardiopulmonary Physical Therapy Journal*, 20(2): 28–31.

Davies, H.W., Teschke, K., Kennedy, S.M., Hodgson, M.R., Hertzman, C. and Demers, P.A. (2005). Occupational exposure to noise and mortality from acute myocardial infarction. *Epidemiology*, 16(1): 25–32.

Din-Dzietham, R., Nembhard, W.N., Collins, R. and Davis, S.K. (2004). Perceived stress following race-based discrimination at work is associated with hypertension in African Americans. The Metro Atlanta Heart Disease Study, 1999–2001. *Social Science and Medicine*, 58(3): 449–461.

Dobie, R.A. (2007). Noise-induced permanent threshold shifts in the occupational noise and hearing survey: an explanation for elevated risk estimates. *Ear and Hearing*, 28(4): 580–591. doi:10.1097/AUD.0b013e31806dc286.

Fang, S.C., Cassidy, A. and Christiani, D.C. (2010). A systematic review of occupational exposure to particulate matter and cardiovascular disease. *International Journal of Environmental Research and Public Health*, 7(4): 1773–1806.

Fernandez, R.C., Driscoll, T.R., Glass, D.C., Vallance, D., Reid, A., Benke, G. and Fritschi, L. (2012). A priority list of occupational carcinogenic agents for preventative action in Australia. *Australia and New Zealand Journal of Public Health*, 36(2): 111–115.

Frost, G. (2013). The latency period of mesothelioma among a cohort of British asbestos workers (1978–2005). *British Journal of Cancer*, 109(7): 1965–1973.

Gawkrodger, D.J. (2004). Occupational skin cancers. *Occupational Medicine*, 54(7): 458–463.

Godderis, L., Mylle, G., Coene, M., Verbeek, C., Viaene, B., Bulterys, S. and Schouteden, M. (2015). Data warehouse for detection of occupational disease in OHS data. *Occupational Medicine*, 65(8): 651–658.

Greiner, B.A., Krause, N., Ragland, D. and Fisher, J.M. (2004). Occupational stressors and hypertension: a multi-method study using observer-based job analysis and self-reports in urban transit operators. *Social Science and Medicine*, 59(5): 1081–1094.

Groenewold, M.R., Masterson, E.A., Themann, C.L. and Davis, R.R. (2014). Do hearing protectors protect hearing? *American Journal of Industrial Medicine*, 57(9): 1001–1010. doi:10.1002/ajim.22323.

Hajakbari, M.S. and Minaei-Bidgoli, B. (2014). A new scoring system for assessing the risk of occupational accidents: a case study using data mining techniques with Iran's Ministry of Labor data. *Journal of Loss Prevention in the Process Industries*, 32(November 2014): 443–453.

Hartley, T.A., Burchfiel, C.M., Fekedulegn, D., Andrew, M.E. and Violanti, J.M. (2011). Health disparities in police officers: comparisons to the U.S. general population. *International Journal of Emergency Mental Health*. 13(4): 211–220.

Humblet, O., Birnbaum, L., Rimm, E., Mittleman, M.A. and Hauser, R. (2008). Dioxins and cardiovascular disease mortality. *Environmental Health Perspectives*, 116(11): 1443–1448.

Hwang, W.J. and Hong, O. (2012). Work-related cardiovascular disease risk factors using a socioecological approach: implications for practice and research. *European Journal of Cardiovascular Nursing*, 11(1): 114–126.

International Labour Organization (ILO) (2010). *ILO List of Occupational Diseases*. www.ilo.org/wcmsp5/groups/public/-ed_protect/-protrav/-safework/documents/publication/wcms_125137.pdf (accessed 3 July 2018).

Kang, M.G., Koh, S.B., Cha, B.S., Park, J.K., Baik, S.K. and Chang, S.J. (2005). Job stress and cardiovascular risk factors in male workers. *Preventive Medicine*, 40(5): 583–588.

Keller, J.E and Howe, H.L. (1993). Cancer in Illinois construction workers: a study. *American Journal of Industrial Medicine*, 24(2): 223–230.

Kim, T.W., Koh, D.H. and Park, C.Y. (2010). Decision tree of occupational lung cancer using classification and regression analysis. *Safety and Health at Work*, 1(2): 140–148.

Kittusamy, N.K. and Buchholz, B. (2004). Whole-body vibration and postural stress among operators of construction equipment: a literature review. *Journal of Safety Research*, 35(3): 255–261.

Knutsson, A., Damber, L. and Jarvholm, B. (2000). Cancers in concrete workers: results of a cohort study of 33,668 workers. *Occupational and Environmental. Medicine*, 57(4): 264–267.

Lacourt, A., Pintos, J., Lavoué, J., Richardson, L. and Siemiatycki, J. (2015). Lunch cancer risk among workers in the construction industry: results from two case-control studies in Montreal. *BMC Public Health*, 15: 941.

Lee, J.H., Kang, W., Yaang, S.R., Choy, N. and Lee, C.R. (2009). Cohort study for the effect of chronic noise exposure on blood pressure among male workers in Busan, Korea. *American Journal of Industrial Medicine*, 52(6): 509–517.

Leensen, M.C.J., van Duivenbooden, J.C. and Dreschler, W.A. (2011). A retrospective analysis of noise-induced hearing loss in the Dutch construction industry. *International Archives of Occupational and Environmental Health*, 84(5): 577–590.

LeMasters, G., Bhattacharya, A., Borton, E. and Mayfield, L. (2006). Functional impairment and quality of life in retired workers of the construction trades. *Experimental Aging Research*, 32(2): 227–242.

Levenstein, S., Smith, M.W. and Kaplan, G.A. (2001). Psychosocial predictors of hypertension in men and women. *Archives of Internal Medicine*, 161(10): 1341–1346.

Lewkowski, K., Heyworth, J., McCausland, K., Fritschi, L., Williams, W. and Li, I. (2017). Predictors of noise exposure in construction workers. *Proceedings of ACOUSTICS 2017*, 19–22 November 2017, Perth, Australia. pp. 1–11. www.researchgate.net/profile/Ian_Li/publication/322306780_Predictors_of_noise_exposure_in_construction_workers/links/5a52fce4aca2725638c7ba86/Predictors-of-noise-exposure-in-construction-workers.pdf (accessed 2 August 2018).

Lin, Y.H., Ho, S.H. and Lai, C.Y. (2015). Physiological workload and musculoskeletal fatigue among construction workers involving squatting/kneeling tasks in Taiwan. *Proceedings of Triennial Congress of the IEA*, Melbourne, 9–14 August 2015. Paper id 129 www.iea.cc/congress/2015/129.pdf.

Lunde, L., Koch, M., Veiersted, K.B., Moen, G., Wærsted, M. and Knardahl, S. (2016). Heavy physical work: cardiovascular load in male construction workers. *International Journal of Environmental Research and Public Health*, 13(4): 356–371.

Masterson, E.A., Tak, S., Themann, C.L., Wall, D.K., Groenewold, M.R., Deddens, J.A. and Calvert, G.M. (2013). Prevalence of hearing loss in the United States by industry. *American Journal of Industrial Medicine*, 56(6): 670–681. doi:10.1002/ajim.22082.

Metcalfe, C., Smith, G.D., Wadsworth, E., Sterne, J.A.C., Heslop, P., Macleod, J., *et al.* (2003). A contemporary validation of the Reeder Stress Inventory. *British Journal of Health Psychology*, 8(1): 83–94.

Miller, B., Fridline, M., Liu, P.Y. and Marino, D. (2014). Use of CHAID decision trees to formulate pathways for the early detection of metabolic syndrome in young adults. *Computational and Mathematical Methods in Medicine.* Article ID 242717.

Mozafari, A., Vahedian, M., Mohebi, S. and Najafi, M. (2015). Work-related musculo-skeletal disorders in truck drivers and official workers. *Acta Medica Iranica*, 53(7): 432–438.

Neitzel, R. and Seixas, N. (2005). The effectiveness of hearing protection among construction workers. *Journal of Occupational and Environmental Hygiene*, 2(4): 227–238. doi:10.1080/15459620590932154.

Ng, D.M. and Jeffery, R.W. (2003). Relationships between perceived stress and health behaviors in a sample of working adults. *Health Psychology*, 22(6): 638–642.

Nicoletti, S. and Battevi, N. (2008). Upper-limb work-related musculoskeletal disorders (UL-WMSDs) and latency of effects. *La Medicina del lavoro*, 99(5): 352–361.

Nowrouzi-Kia, B., Li, A.K.C., Nguyen, C. and Casole, J. (2018). Heart disease and occupational risk factors in the Canadian population: an exploratory study using the Canadian Community Health Survey. *Safety and Health at Work*, 9(2): 144–148.

Oude Hengel, K.M., Blatter, B.M., Geuskens, G.A., Koppes, L.L. and Bongers, P.M. (2012). Factors associated with the ability and willingness to continue working until the age of 65 in construction workers. *International Archives of Occupational and Environmental Health*, 85(7): 783–790.

Pope, C.A., Burnett, R.T., Thurston, G.D., Thun, M.J., Calle, E.E., Krewski, D. and Godleski, J.J. (2004). Cardiovascular mortality and long-term exposure to particulate air pollution: epidemiological evidence of general pathophysiological pathways of disease. *Circulation*, 109(1): 71–77.

Prokopowicz, A., Sobczak, A., Szuła-Chraplewska, M., Zaciera, M., Kurek, J. and Szołtysek-Bołdys. I. (2017). Effect of occupational exposure to lead on new risk factors for cardiovascular diseases. *Occupational & Environmental Medicine*, 74(5): 366–373.

Pukkala, E., Soderholm, A.L. and Lindqvist, C. (1994). Cancers of the lip and oropharynx in different social and occupational groups in Finland. *European Journal of Cancer. Part B, Oral Oncology*, 30B(3): 209–215.

Radi, S., Lang, T., Lauwers-Cances, V., Diene, E., Chatellier, G., Larabi, L., *et al.* (2005). Job constraints and arterial hypertension: different effects in men and women: The IHPAF II case control study. *Occupational & Environmental Medicine*, 62(10): 711–717.

Ramaswami, M. and Bhaskaran, R. (2010). A CHAID based performance prediction model in educational data mining. *International Journal of Computer Science Issues*, 7(1): 10–18.

Reid, A., de Klerk, N.H., Magnani, C., Ferrante, D., Berry, G., Musk, A.W. and Merler, E. (2014). Mesothelioma risk after 40 years since first exposure to asbestos: a pooled analysis. *Thorax*, 69(9): 843–850.

Riva, M.M., Bancone, C., Bigoni, F., Bresciani, M., Santini, M. and Mosconi, G. (2012). Work-related diseases and the fitness to work in the construction industry. *Giornale italiano di medicina del lavoro ed ergonomia*, 34(3): 306–312.

Safe Work Australia (2014). *Occupational Disease Indicators.* www.safeworkaustralia.gov.au/system/files/documents/1702/occupational-disease-indicators-2014.pdf (accessed 19 October 2018).

Safe Work Australia (2015). *Deemed Diseases in Australia.* www.safeworkaustralia.gov.au/system/files/documents/1702/deemed-diseases.pdf (accessed 12 March 2018).

Safe Work Australia (2016). *The Australian Work Exposures Study (AWES) – Carcinogen Exposures in the Construction Industry.* www.safeworkaustralia.gov.au/system/files/documents/1702/awes_-_carcinogen_exposures_in_the_construction_industry.pdf (accessed 11 March 2019).

Safe Work Australia (2017). *Work-related Disease Research.* www.safeworkaustralia.gov. au/statistics-and-research/research-and-studies/work-related-disease-research (accessed 18 January 2018).

Seixas, N.S., Goldman, B., Sheppard, L., Neitzel, R., Norton, S. and Kujawa, S.G. (2005). Prospective noise induced changes to hearing among construction industry apprentices. *Occupational and Environmental Medicine,* 62(5): 309–317. doi:10.1136/ oem.2004.018143.

Sobeih, T.M., Salem, O., Daraiseh, N., Genaidy, A. and Shell, R. (2006). Psychosocial factors and musculoskeletal disorders in the construction industry: a systematic review. *Theoretical Issues in Ergonomics Science,* 7(3): 329–344.

Steenland, K., Fine, L., Belkić, K., Landsbergis, P., Schnall, P., Baker, D., Theorell, T., Siegrist, J., Peter, R., Karasek, R., Marmot, M., Brisson, C. and Tüchsen, F. (2000). Research findings linking workplace factors to CVD outcomes. *Occupational Medicine,* 15(1): 7–68.

Steptoe, A., Siegrist, J., Kirschbaum, C. and Marmot, M. (2004). Effort-reward imbalance, overcommittment, and measures of cortisol and blood pressure over the working day. *Psychosomatic Medicine,* 66: 323–329.

Theorell, T., Alfredsson, L., Westerholm, P. and Falck, B. (2000). Coping with unfair treatment at work – what is the relationship between coping and hypertension in middleaged men and women? *Psychotherapy and Psychosomatics,* 69(2): 86–94.

Tomei, G., Fioravanti, M., Cerratti, D., Sancini, A., Tomao, E., Rosati, M.V., Vacca, D., Palitti, T., Di Famiani, M., Giubilati, R., De Sio, S. and Tomei, F. (2010). Occupational exposure to noise and the cardiovascular system: a meta-analysis. *Science of the Total Environment,* 408(4): 681–689.

University of Michigan Health System (2014). *What is Ischemic Heart Disease and Stroke.* www.med.umich.edu/1info/FHP/practiceguides/cad/IHDshort.pdf (accessed 5 July 2018).

University of Ottawa Heart Institute (2011). *Coronary Artery Disease: A Guide for Patients and Families.* www.ottawaheart.ca/sites/default/files/uploads/coronary-artery-disease-patient-guide.pdf (accessed 5 July 2018).

Wang, X., Dong, X.S, Choi, S.D. and Dement, J. (2016). Work-related musculoskeletal disorders among construction workers in the United States from 1992 to 2014. *Occupational and Environmental Medicine,* 74(5): 374–380. http://dx.doi.org/10.1136/oemed-2016-103943.

Yang, H., Schnall, P.L., Jaurequi, M., Su, T.C. and Baker, D. (2006). Work hours and self-reported hypertension among working people in California. *Hypertension,* 48(4): 744–750.

Zhang, J., Liu, Y., Shi, J., Larson, D.F. and Watson, R.R. (2002). Side-stream cigarette smoke induces dose-response in systemic inflammatory cytokine production and oxidative stress. *Experimental Biology and Medicine,* 227(9): 823–829.

4 Curbing psychological injuries in construction using analytics

Introduction

The construction industry is characterised by one of the worst safety performances of any industry globally. The unacceptably high accident rates in construction cause not only human suffering but also productivity losses, project delays, increased project costs and damage to the reputation of the builder (Fung *et al.* 2009; Gangolells *et al.* 2010). Because of these grave consequences, safety has been a hot subject for research and extensive studies have been conducted on various topics, among them are accident causation modelling, workers' compensation for accidents, return-to-work and cost of accidents.

Construction is equally recognised for high levels of work stress and related psychological injuries. Psychological injury is a broad term that refers to any form of mental ill-health caused by work stress and includes: anxiety, depression, mood disorder, substance use, suicidality, etc. Because psychological risks suffered by construction operatives and professionals are invisible and silent, unlike physical injuries, they tend to go unnoticed for extended periods, causing serious damage to individuals, their families and society as a whole. Yet the psychological well-being of construction operatives and professionals is an underexplored subject in the construction domain. To that end, this chapter aims to address the following research question by applying data mining and analytics methods:

> What are the association patterns and causes of psychological injuries in the Australian construction industry?

Psychological injuries are the consequence of excessive, prolonged work stress. Hence, the chapter begins by defining work stress, its effects and theories of work stress. Then, the overall research method adopted and the specific analytics technique used for knowledge discovery are discussed. Following that, data mining of psychological injury data and findings are presented. Finally, conclusions are drawn whereby suggestions are offered for mitigating the problem and for further research.

Work stress

Work stress is defined variably by different bodies; however, these definitions provide similar explanations to the concept. For instance, the National Institute of Occupational Health and Safety (1999) in the US defined work stress as 'the harmful physical and emotional responses that occur when the requirements of the job do not match the capabilities, resources or needs of the worker. Similarly, the Health and Safety Executive (2001) in the UK stated that 'work stress is the reaction people have to excessive pressure or other types of demand placed on them'. WorkCover NSW (2014) in Australia described work stress as

> the physical, mental and emotional reactions of workers who perceive that their work demands exceed their abilities and/or their resources (such as time, skills, support) to do the work. It occurs when they perceive they are not coping with situations where it is important to them that they cope.

Leung *et al.* (2015, p. 91) argued that stressors are an essential component of stress. In the work context, stressors are defined as stimuli individuals face on the job that have negative physical and psychological effects on a significant portion of people exposed to them (Ganster and Rosen 2013). These are aspects related to work design, the organisation and management of work, which have the potential to cause psychological and physical harm (also known as psychological hazards) (Cox *et al.* 2005).

It can be derived from a critical look at the above definitions and the concept of stressors that four key constructs constitute work stress, namely: stressors, mental strain/pressure, coping and effects. Hence, it can be conceptualised that:

> Work stress is the result of the failure of individuals to cope with the mental strain/pressure exerted by stressors at work and it affects their psychological and physical well-beings.

It can be further articulated that an optimum level of strain/pressure at work, if managed effectively, can be a stimulus that motivates workers and drives performance. Stress materialises only when the strain/pressure is continuously intense, excessive and/or not managed appropriately.

Effects of work stress on individuals and organisations

Work stress can have many effects on the health of individuals and these effects may take the form of physiological and psychological responses over time (Cox *et al.* 1983). The result of these burdens may have a consequential effect on their work performance and productivity of individuals, leading to larger implications on organisational operations (Dewe *et al.* 2000). Figure 4.1 summarises the multifaceted effects of work stress and the following sections explain them individually.

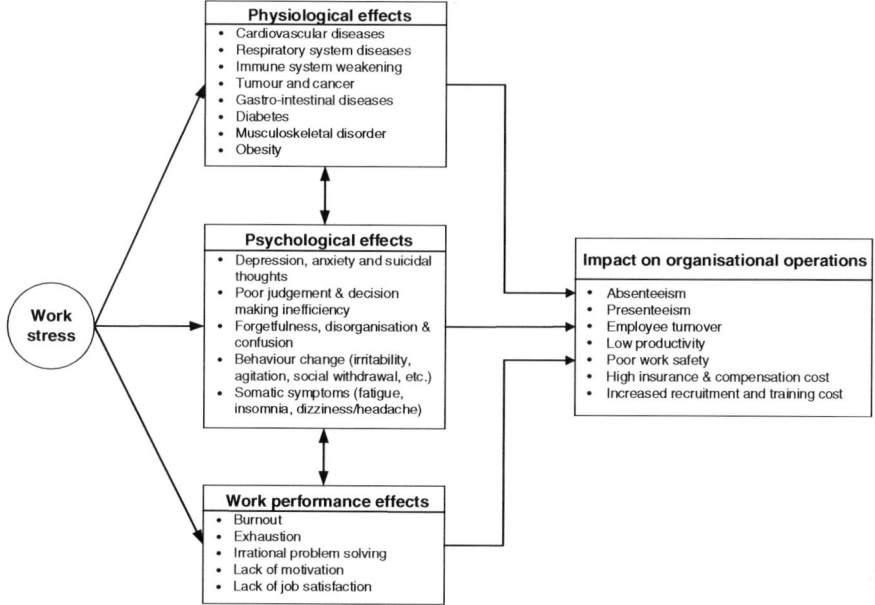

Figure 4.1 Effects of work stress.

Physiological effects

Physiological effects refer to impacts on an individual's physical health that may be detrimental to healthy or normal functioning (Kawakami and Haratani 1999).

Medical studies have found that there is an elevated risk of cardiovascular diseases when individuals experienced job strain due to high job demands and low decision-making opportunities (Jex 1998; Belkić *et al.* 2000; Marshall *et al.* 2001). Similarly, Dimsdale (2008) suggested that chronic stress (i.e. continuous exposure to stress) increases the risk of cardiac events (e.g. heart attacks) and cardiac diseases. Steptoe and Kivimäki (2012) reinforced this, stating that continued exposure to chronic and acute occupational stress could lead to increased risk of recurrent coronary heart disease events as well as mortality.

Work stress has also been found to affect the respiratory systems of individuals (Cox *et al.* 2000). Jacobs *et al.* (1970) performed a study on 179 male college students and found that stress was associated with the presence of respiratory illnesses such as asthma, upper-respiratory infection and rhinitis. Cohen and Williamson (1991) maintained that stress was a potential reason for upper respiratory inflammation caused by bacterial infections, producing symptoms such as sore throats, nasal congestion and mucus secretion.

Effects of work stress on the immune system have also been reported in many studies. Godbout and Glaser (2006) argued that stress weakens the immune system, which leads to delays in wound healing processes and increases the severity and duration of infectious diseases. Stress had also been suggested to have an effect on tumour promotion and cancer, as stress may result in a reduction in the activity of natural killer and cytotoxic cells that can target abnormally growing cells for destruction (Glaser *et al.* 2005).

There has also been research linking work stress with problems in the gastrointestinal system (Cox 1993; Cox *et al.* 2000). Räihä *et al.* (1998) performed a questionnaire survey on 13,888 twins from Finland and their logistic regression analysis found that high stress levels from daily activities was a significant risk factor for peptic ulcer disease (a lesion in the lining of the digestive tract). Stanghellini (1998) conducted a questionnaire survey of 5581 individuals internationally and found that psychological stress was a notable risk factor for upper gastro-intestinal symptoms (such as stomach aches and indigestion). Chen *et al.* (2009) performed a cross-sectional survey of 561 Chinese male offshore oil workers and found that there was a positive association with work stress and ulcer-like symptoms (defined as localised epigastric pain, hunger pain and nocturnal gastric pain). Stress has also been reported to have an effect on the female reproductive system (Kalantaridou *et al.* 2004).

There are also other non-direct physiological effects resulting from work stress. These are primarily caused by changes in the behaviour of the individual, for example changes in eating habits causing obesity, sleep deprivation causing elevated insulin, blood glucose and pressure levels, poor work ergonomics causing musculoskeletal disorders and the consequences from drinking alcohol and smoking (Bongers *et al.* 1993; Carayon *et al.* 1999; McEwen 2006, 2007, 2008).

Psychological effects

Dollard (2001) categorised the psychological changes resulting from work stress into four parts: (i) cognitive effects; (ii) behaviour effects; (iii) affective disorders; and (iv) somatic symptoms. In terms of cognitive effects, Flier *et al.* (1998) suggested that repeated stress affects the hippocampus (the central part of brain function affecting emotions and memory), which in turn has the potential to affect verbal and short-term memory (where verbal memory refers to memory of context related to events). Eysenck *et al.* (2007) suggested that stress affects a person's processing efficiency as it reduces attentional control due to threat-related distracting stimuli (stress). They further claimed that an individual's inability to concentrate was attributed to task-irrelevant stimuli that was internal (e.g. worrying thoughts) as well as external (conventional distracters).

The behaviour of individuals has been suggested to change as a result of work stress (Cooper *et al.* 1996; Cox *et al.* 2000; Dollard 2001). Lindquist *et al.* (1997) explained that changes in behaviour were strategies (often harmful) that individuals used in order to cope with stress, and included alcohol and drug abuse, smoking, binge eating and interpersonal withdrawal. Emotion and stress have also been suggested to share many characteristics, and Hart and Cooper (2001) claimed that unpleasant emotions such as tension, frustration and anger were also the result of work stress.

In terms of affective disorders, Lupien *et al.* (2009) suggested that the presence of high levels of glucocorticoids (resulting from chronic and/or acute stress) affected the frontal cortex and hippocampus of adults and had the potential to increase the risk of depressive disorders. Rada and Johnson-Leong (2004) explained that depression affects the body, mood and thoughts that change the way in which individuals felt about themselves. In addition to major depressive disorders, stressful work conditions have also been suggested to cause generalised anxiety disorders in individuals (Melchior *et al.* 2007). In some extreme cases, it has also been suggested that long-term psychological effects of stress may result in mental illness or even suicide (Dollard 2001).

Work stress has also been suggested to cause somatic symptoms (involving pain or discomfort) such as nausea, headache, upset stomachs (Spector and Jex 1998), perspiration and dizziness (Dollard 2001). Nomura *et al.* (2007) conducted a cross-sectional study of 185 Japanese male office workers (between 21 and 66 years old) and showed that the most frequent somatic symptoms experienced (resulting from work stress) included general fatigue, sleep disturbance and headache.

Work performance effects on individuals

Work performance can be defined as the ability of an individual to successfully accomplish an assigned task or objective with a reasonable utilisation of available resources (Jamal 1984). Work stress is suggested to have varying effects on the work performance of individuals. The inverted U-shaped relationship between stress and work performance suggests that at low levels of stress, individuals are not sufficiently aroused for high performance, whereas at high levels of stress, individuals expend their energy coping with stresses rather than achieving the assigned task (Abramis 1994). Abramis (1994) further explained that at moderate (or optimal) levels of stress, individuals were activated and motivated for high performance in their role (hence the inverted U-shaped relationship). Leung *et al.* (2008) conducted a questionnaire survey with 108 construction project managers to establish the association between objective stress (individual's ability to carry out a task) and interpersonal performance, as well as between burnout, physiological stress and organisational performance. They suggested that when stress was low, under-stimulation caused boredom and fatigue, whilst with high stress, over-stimulation caused irrational problem solving and exhaustion. For moderate amounts of stress however, they noted that employees would be in the optimum stimulation zone (with high performance) where rational problem solving, creativity and satisfaction occurred.

Impacts of work stress on organisational operations

Cox *et al.* (2000) argued that when a large number of employees are experiencing the effects of stress in their workplace, then stress is said to be of organisational proportions. They further suggested that if 40 per cent of workers in any department of an organisation were facing stress-related issues, then the organisation was considered unhealthy or in ill-health. Cooper *et al.* (1996) suggested that there were a wide range of indices that are indicative of organisational ill-health including sickness absence, high labour turnover and low productivity. They further suggested that organisational ill-health may lead to issues such as poor work safety records, increased recruitment and training costs, low levels of organisational commitment, low levels of job satisfaction and deteriorating industrial and customer relations. This was supported by Cox *et al.* (2000) who suggested that the most frequently cited organisational issues included reduced availability for work (high turnover), absenteeism, presenteeism and impaired work productivity. Direct financial impacts can include high insurance and healthcare costs (Cooper *et al.* 1996; Manning *et al.* 1996) and/or increases in employee compensation claims (Cox *et al.* 2000; Macklin *et al.* 2006; Keegel *et al.* 2009).

Theories of work stress

Many theories of work stress have so far been postulated by psychological theorists. These can be grouped under three broad categories, namely: interactional theories, transactional theories and multi-perspective theories. The interactional theories deal with the structural aspects of an individual's interaction with the work environment (i.e. focusing on the architecture of situations that result in the experience of stress) (Cox *et al.* 2000). The transactional theories, on the other hand, focus on the processes involved in the experience of stress such as cognitive appraisal, emotional reactions and coping ability (Cox and Griffiths 2010). The multi-perspective theories build upon specific concepts related to the interactional or transactional theories of work stress (Dewe *et al.* 2012). Table 4.1 summarises the sources of work stress as postulated by the theories. Each theory tries to explain the source of work stress form a particular angle and has its own list of stressors.

Table 4.2 collates all the different stressors from the theories and organises them under three key factors: personal, work and socio-economic. The combined list provides a broader picture of the causes of work stress and is used to guide the data mining and analytics exercise related to the psychological injuries suffered by construction operatives and professionals.

Methods and materials

The above review of theories summarises the various causes of work stress. However, how work factors, along with socio-economic and personal factors collectively influence the severity of psychological injuries suffered by victims is yet to be explored. This study utilised the Multiple Correspondence Analysis (MCA) technique, which is the extension of Correspondence Analysis (CA), for discovering the association patterns among the said factors using past psychological injury data. Further details about the benefits of using MCA are discussed in Chapter 2.

Table 4.1 Work stress theories and premises

Theory	Premise of the theory
Interactional theories	
Person–Environment Fit Model (French *et al.* 1982)	Stress arises from the misfit between the person and the environment; that is, because of a misfit between the abilities of the employee and the demands of the jobs or between the needs of the employee and the rewards from the job.
Demand–Control–Support (DSC) Model (Karasek and Theorell 1990)	Stress occurs by the combined effect of work demand, control and support. A combination of high job demand and low levels of job control, particularly decision making freedom, and low support from superiors or co-workers would lead to 'high-strain, isolated jobs'. On the contrary, high job demands combined with high levels of control and support would lead to 'active jobs' that are not stressful as they allow individuals to regulate job demands.
Transactional theories	
Effort–Reward Imbalance (ERI) Model (Siegrist 1996, 2009)	Work stress is triggered when an imbalance is perceived by an employee between the efforts (job demand, obligations, time pressure, overtime, and performance expectations) he/she puts in and the rewards (pay, esteem, job security, career development) received.
Appraisal and Coping Model (Lazarus and Folkman 1984; Cox *et al.* 2000; Miller and McCool 2003)	Individuals' appraisal of situations and their coping ability is central to stress arousal, rather than the external environment/event itself. Different individuals appraise the same environment/event differently, depending on their experience and psychological characteristics. Stress occurs when a person appraises a situation to be harmful and doubt their ability to cope with it.
Multi-perspective theories	
Vitamin Model (Warr 1987)	The theory argues that job characteristics and affective well-being pose a curvilinear relationship. Job characteristics (e.g. job demands, job decision latitude, social support, utilisation of abilities, salary, safety and work task significance) have an initial beneficial effect on employee's mental health and, beyond a certain required level, may produce either a constant effect on health (similar to vitamins C and E) or may be harmful to health (similar to vitamins A and D).

continued

Table 4.1 Continued

Theory	Premise of the theory
Organisational Health Framework (Hart and Cooper 2001)	Employee well-being (morale, distress and job satisfaction) is affected by a dynamic system of interactions between multiple individual characteristics (personalities, coping processes, attitude and behaviours) and organisational characteristics (resources, structure, culture and organisational climate), which, in turn, affect the organisation's performance (company turnover, absence, medical expenses, customer satisfaction and compensation claims).
Cooper and Marshall's Model (Cooper and Marshall 1976)	Work stress is the outcome of interplay between work factors (intrinsic to the job, role in the organisation, relationships at work, career development, organisational structure and climate, etc.), personal life circumstances (family problems, life crises, financial difficulties, etc.) and individual characteristics (Type A personality, levels of anxiety, neuroticism, tolerance and resilience, etc.).
Demand Induced Strain Compensation (DISC) Model (de Jonge and Dormann 2003)	Work stress results from a lack of job resources to address job demands. Adverse health effects of high job demands could be compensated for, by matching job resources (physical resources, social/emotional support from colleagues, breaks, availability of control, etc.) to the high demands. An unbalanced mix of specific job demands and job resources activates psychological compensation processes of strains whilst a well-balanced mix of specific job demands and job resources could stimulate employee growth and performance.
Cybernetics Model (Edwards 1992)	The basic premise of the theory is that individuals are the managers of stress and they seek to maintain an equilibrium state and attempt to re-establish equilibrium when an external force breaks it. Discrepancy between perceptions (the subjective representation of any situation, condition, or event) and desires (work goals and interests) is the cause of stress. Depending on the importance of the discrepancy, the stress may lead to an alteration of well-being and/or the employment of coping strategies.
Dynamic Equilibrium Model (Headey and Wearing 1989; Headey 2006)	Stress arises when there is a state of disequilibrium that affects an individual's normal levels of psychological distress and well-being. Stress is suggested to come from a system of variables including personality, environmental characteristics, coping processes, positive and negative experiences as well as indices of psychological well-being.

Ethological Theory (Schabracq *et al.* 2003)	Employees have their own 'territory' (i.e. a space of familiarity resulting from routine actions) where they develop standardised behavioural sequences and knowledge needed for survival. Infractions or loss of one's territory lead to stress and, consequently, the activation of self-defence mechanisms (including identification of the problem and evaluation of coping strategies).
Individual Difference Factors Model (Bright 2001)	Individuals respond uniquely to stressors depending on the genetic, dispositional and acquired differences. Genetic factors concern age, gender and physique. Dispositional factors are self-esteem, coping style and negative affectivity. Examples for acquired factors include social class, education, job position and social support.
Multidimensional Theory of Burnout (Maslach 1998)	High job demands (work overload and personal conflicts) and a lack of resources (social support, skill use, autonomy, decision involvement and control) contribute to the development of occupational burnout (exhaustion, detachment from the job, frustration, anger and feelings of ineffectiveness and failure), which in turn, affect both the employee and his/her organisation (diminished organisational commitment, turnover, absenteeism and physical illness).
Job Demands – Resources (JDR) Model (Demerouti *et al.* 2001; Bakker and Demerouti 2007)	Interactions between job demands and job resources affect the level of strain and motivation in employees. Job demands are physical, social or organisational components of a job that require sustained physical and/or mental effort. Job resources necessary to meet the job demand include pay, career opportunities, supervisor support, workplace culture, decision-making latitude, clearly defined role, task significance, autonomy and performance feedback. High job demand compensated by high job resources will lead to high motivation and average (manageable) strain in employees).
Demands, Resources and Individual Effects (DRIVE) Model (Mark and Smith 2008)	Work demands, individual differences and work resources have a main effect relationship on health outcomes and job satisfaction. Whilst work resources and individual differences moderate the relationship between work demands and health outcomes, perceived job stress is a mediator for work demands, resources, and health outcomes (along with job satisfaction).

Table 4.2 Work stress factors

Stress source	Stressor
Personal factors	Demographics (age, gender, education, occupation) Personality (Type A behaviour pattern) Coping and appraisal process Psychological characteristics (level of: anxiety, neuroticism, tolerance) Previous exposure to stress
Work factors	*Job demand*: role and responsibilities, workload, time pressure, physical workplace. *Job resources*: physical resources, social relationships and support, autonomy, performance feedback. *Organisational characteristics*: structure, decision making culture, office politics. *Rewards*: pay, esteem, job security, career development.
Socio-economic factors	Family relationships and support, other social relationships and support, life crises, economic circumstances.

Data

Psychological injury data required for the research was obtained in March 2016 from Safe Work Australia, which is a government agency responsible for leading the development of national policy to improve work health and safety. It compiles work-related compensation claims and fatalities datasets of states and territories, including workplace incidents across all industries. When a formal request for construction accident and compensation data was made to the federal body by the author, they collated data from the different jurisdictional workers' compensation authorities and provided them to the author after the formal signing a confidentiality agreement. The entire process took about 12 months, but the dataset was large, and encompassed 391,494 cases of workers' compensation claims filed by the construction industry across Australia over a 13-year period, between 2002 and 2014. Filtering the database, a subset of 3898 cases of approved psychological injury claims was extracted for this study. A typical case was characterised by 23 variables and the definitions of these variables are explained in Table 2.1.

Data preparation

Pre-processing was undertaken to prepare the data for mining. First, 28 cases that are not related to the construction industry, but had been incorrectly recorded, were removed from the dataset. Then, further checking was performed to identify cases that are not significant for analysis. Two-hundred and eighty-six cases with both zero lost days and zero compensation were identified and removed. Another checking resulted in removing a further 156 cases that had zero compensation, resulting in 3428 cases. A final filtering was undertaken to remove cases with zero lost days, resulting in 2268 usable cases for data mining.

Table 4.3 Selected variables for data mining

Variable	Measurement	Classification system followed
Age	Under 20; 20 to 29; 30 to 39; 40 to 49; 50 to 59; 60 and above	Derived from ABS classification
Size of employer	Micro (up to 4 FTE); small (5–19 FTE); medium (20–199 FTE); large (200 or more FTE)	Derived from ABS classification
Gender	Male; Female	Natural
Occupation	Managers; Professionals; Technicians and trade workers; Community and personal service workers; Clerical and administrative workers; Sales workers; Machinery operators and drivers; labourers	ANZSCO classification
Hours usually worked weekly	Less than 20 hours; 20 to 38 hours; 39 to 49 hours; 50 hours or more	Derived from ABS classification
Normal weekly earnings	Less than $385; $385 to $769; $770 to $1154; $1155 to $1539; $1540 to $1924; $1925 to $2885; Above $2885	Derived from ABS classification
Mechanism of psychological injury	Adjustment disorder; assault; exposure to traumatic events; exposure to workplace violence; poor physical work environment; post-traumatic stress disorder; sexual or racial harassment; suicide or attempted suicide; work pressure; workplace harassment and/or bullying; other mental stress factors (dietary or deficiency diseases – Bulimia, Anorexia); multiple stress factors; unspecified stress factor	TOOCS3.0 classification
Psychological injury severity	Minor; moderate; major; severe; critical	Previous research

Upon data cleaning, the dataset was redefined with eight variables for multiple correspondence analysis. The other attributes that described incident circumstances were not removed nor were they used for data mining. Table 4.3 illustrates the variables used for data mining. All the variables were made categorical to enable non-parametric data mining with multiple correspondence analysis. The categories used for variables were derived from different classification systems available in Australia. The mechanisms of injury, for instance, were based on the Type Of Occurrences Classification System (TOOCS) Third Edition of the Australian Safety and Compensation Council. Similarly, occupation classifications were based on the Australian and New Zealand Standard Classification of Occupations (ANZSCO) First Edition. Grouping intervals for age, weekly earning and work hours were derived from the classifications used by the Australian Bureau of Statistics (ABS) in its various reports.

Categories for the severity of psychological injuries were based upon the lost time reported for the cases. The utilisation of lost time to categorise injuries into different severity levels has commonly been used by previous researchers who investigated physical injuries in construction; for example, Dumrak *et al.* (2013) and Arquillos *et al.* (2012). Nonetheless, varied severity levels and groupings of lost days were implemented by different researchers. This research adopted five levels of severity such as minor, moderate, major, sever and critical, and the definitions for these categories are given below, derived from Kamardeen and Rameezdeen (2016, p. 621) who had developed them based on a risk matrix recommended by the National Patient Safety Agency (NPSA) (2008).

- **Minor** – psychological injuries requiring time off work (lost days) for fewer than three days.
- **Moderate** – psychological injuries requiring time off work (lost days) for four to 14 days.
- **Major** – psychological injuries requiring time off work (lost days) for 15 to 99 days.
- **Severe** – psychological injuries requiring time off work (lost days) for 100 to 365 days.
- **Critical** – psychological injuries requiring time off work (lost days) for more than 365 days.

Days were used as the scale of measurement of lost time for categorising the severities. Because lost time in the dataset used in this research had been recorded in hours, equivalent hours were used to re-interpret these severity categories, based on a 38-hour week of five workdays, producing 7.6 hours per workday.

Data mining discoveries and discussions

The data mining and analytics exercise aimed to answer the following two questions:

- What are the significant combination groups that result in serious psychological injuries in the construction industry?
- Are there any associations among the categories of variables in explaining the categories of psychological injury severities?

IBM SPSS version 24 was deployed for data mining. Data optimisation in MCA can be performed with multiple dimensions (axes); however, Greenacre and Blasius (2006) suggested that a two-dimensional representation is adequate to explain the majority of variances in MCA. Moreover, Das and Sun (2015) suggested that morphological maps are an effective way of presenting information visually as it allows one to interpret the distribution of variable category combinations. Hence, the MCA analysis for this study was performed with two dimensions and the graphical results are presented below, which include: model summary, object plot and joint plot of category points.

The mode summary, displayed by Table 4.4, reveals that 46.4 per cent of the records optimally present the dataset in the two-dimensional space, as explained by the total inertia. Each of the dimensions has yielded an eigenvalue higher than 1.00, which is the threshold value to determine whether to include a dimension for data reduction and optimisations.

The object plot illustrated in Figure 4.2 displays the position of each data point in the two-dimensional space. It is apparent from the plot that no outlier is present in the dataset that is used for knowledge discovery and therefore the results are noise-free. There is a good spread of cases along the dimensions with the majority located between –0.5 and 1.00 on dimension 1 and –1.5 and 2.0 on dimension 2; close to the origin. This supports the decision to use only the two-dimensions for MCA, although it seems the total inertia for both dimensions is only around half of the total data variability detected. Nonetheless, the two dimensions comprehensively capture the total variability present in the dataset.

The most important visual output is the joint plot of category points, which maps the distribution of categories of variables with dimension coordinates. Figure 4.3 illustrates the complete joint plot for the study and several insights can be derived by examining it. Specifically, the scrutiny of combination clouds;

Table 4.4 Model summary

Dimension	Cronbach's alpha	Variance accounted for		
		Total (eigenvalue)	Inertia	% of variance
1	0.542	1.901	0.238	23.760
2	0.511	1.809	0.226	22.614
Total		3.710	0.464	
Mean	0.527[a]	1.855	0.232	23.187

Note
a Mean Cronbach's alpha is based on the mean eigenvalue.

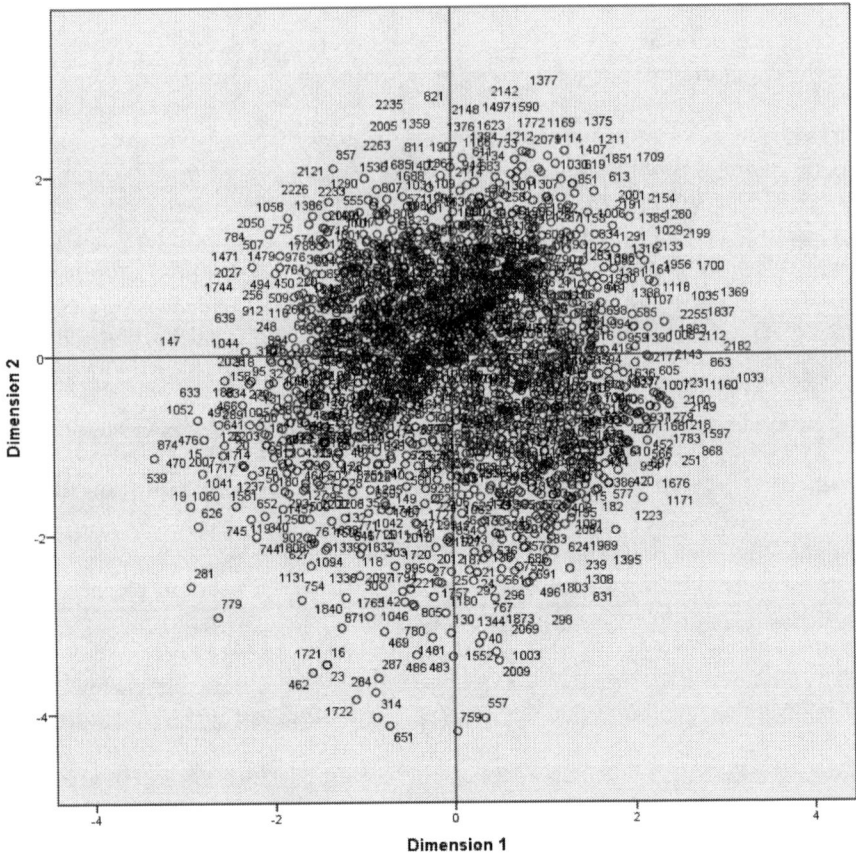

Figure 4.2 Object plot.

i.e. categories that are relatively close form combinations. Five individual combination clouds are derived from the joint plot and are explained below.

Combination cloud 1 (Figure 4.4) reveals that work pressure, workplace harassment and/or bullying and poor physical work environment are closely associated
with more serious psychological injuries in construction, which are categorised as
major, sever and critical. Moreover, it largely affects professionals in the construction industry in almost all sizes of organisations. Research conducted by Bowen
et al. (2014a) in the South African construction industry found five predictors of
work stress among construction professionals, including: work-family-life imbalance, the need to prove one's value, hours worked per week, working to tight deadlines and lack of support from the line manager in difficult situations. The latter
three stressors constitute work pressure. Similarly, in relationship to the Victorian
construction industry in Australia, Haynes and Love (2004) sorted the top three

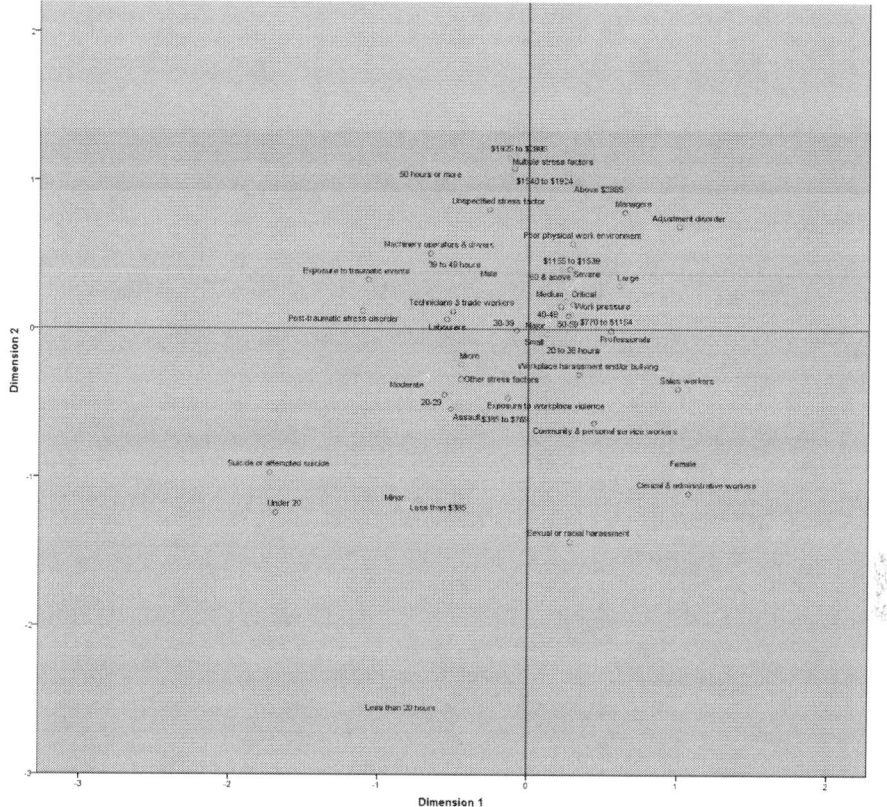

Figure 4.3 Joint category plot.

stressors for site managers, namely workload (constant time pressure), working long hours and insufficient time spent in family/home environment. Lingard and Francis (2004) found that site-based employees suffer significantly higher levels of job-related emotional exhaustion and work interference with family life than office-based employees because they work more hours. They further argued that, on average, site-based employees work over 60 hours a week whilst office-based employees work 40–50 hours. Similarly, Boschman *et al.* (2013) found that high work speed, high volume of work, low participation in decision-making, and low support from the direct supervisor are associated with symptoms of depression among construction supervisors. Issues of work pressure, harassment and/or bullying and poor physical environment affect not only the mental well-being but also physical safety of construction workers. Leung *et al.* (2010) posited that construction incident rates are influenced by emotional stress of workers, which is predicted by work overload, inter-role conflict, poor physical environment, unfair treatment, and inappropriate safety equipment.

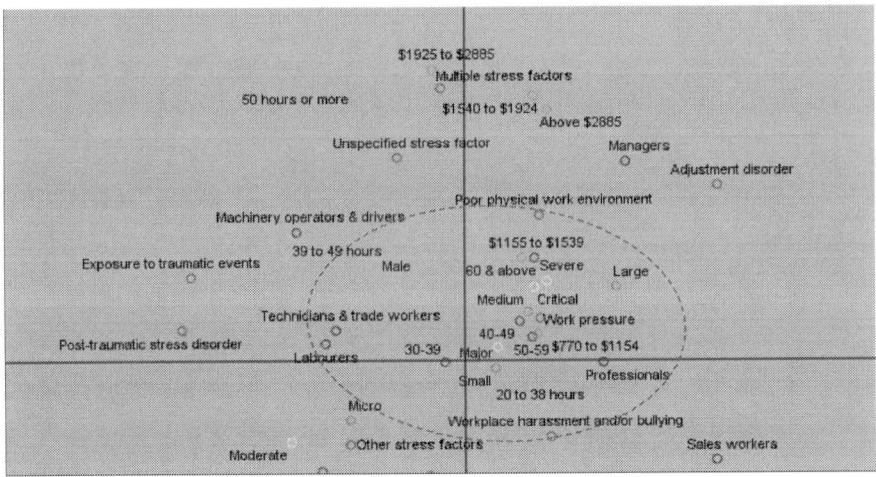

Figure 4.4 Combination cloud 1.

Bowen *et al.* (2014b) pointed out that female professionals working on construction sites experience issues related to the provision, adequacy and cleanliness of female toilet amenities. Poor physical work environment, which has been discovered in this study, represents this issue. However, in the Australian context, this is not a gender-specific issue. Bowen *et al.* (2014b) further found that harassment is a significant stressor in the construction industry, which covers: harassment of a sexual nature (unwanted suggestion about or reference to sexual activity, unwanted physical contact and unwanted physical contact of a sexual nature) and harassment (threatening verbal or physical conduct or exclusionary behaviour) because of language, ethnicity, religion and/or gender. Female professionals mentioned that male artisans were less willing to accept instructions from female supervisors. Likewise, discrimination is a significant stressor too in the construction industry, which manifests as unequal treatment by line managers because of ethnicity, language, religion or gender. At workers level, Goldenhar *et al.* (1998) reported that psychological well-being of female construction workers is impacted by sexual harassment, gender discrimination, job uncertainty and over compensation at work.

Combination cloud 2 (Figure 4.5) indicates that those in managerial positions in the construction industry are affected by multiple stress factors and adjustment disorders, although they do not face income-related stress. Poor physical work environment seems to affect them too, although this is more relevant to site-based managers. Haynes and Love (2004) concurred with this combination cloud that less experienced managers are at a greater risk of adjustment problems than their more experienced counterparts. Sutherland and Davidson (1993) identified ten key stressors simultaneously encountered by construction site managers in the UK, which could fall under the label 'multiple

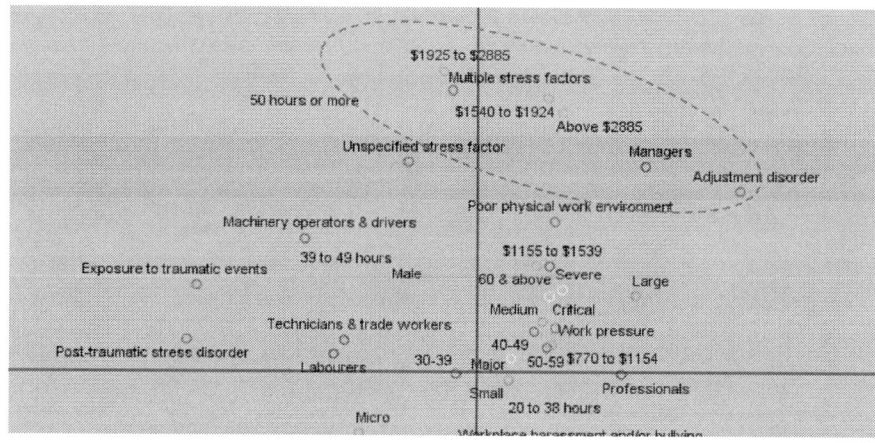

Figure 4.5 Combination cloud 2.

stress factors' in the combination cloud. These stressors are: time pressure, working long hours, volume of paper work, staff shortage, responsibility for situation not fully in one's control, insufficient time to pursue leisure interests, insufficient time spent with family/home, travel to and from the job, lack of support from architects and inadequacy of communication flow.

Combination cloud 3 (Figure 4.6) concerns technicians, trade workers, labourers, machinery operators and drivers. Machinery operators and drivers suffer mental ill-health largely due to being exposed to traumatic events, which

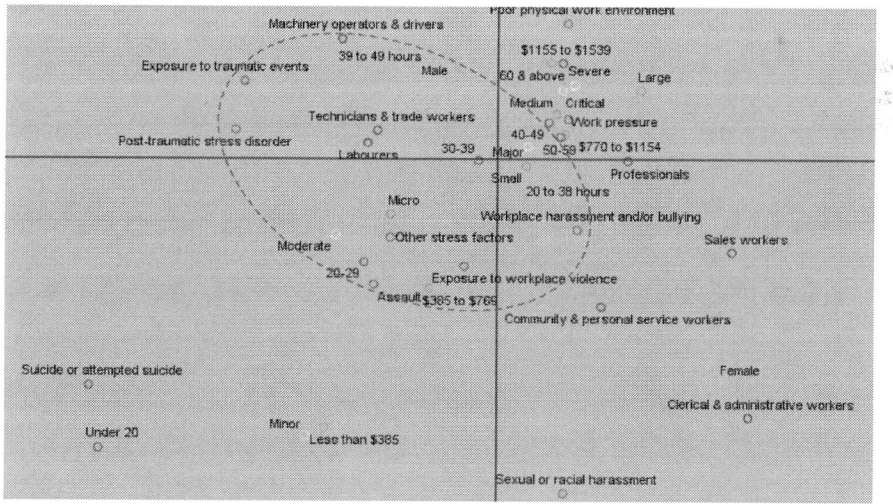

Figure 4.6 Combination cloud 3.

are likely to be sudden vehicular collisions or toppling of machinery they operate. This could also lead to post-traumatic stress disorder. The level of psychological ill-health suffered by this group is between severe and critical. When exposed to traumatic events, technicians, trade workers and labourers also develop subsequent post-traumatic stress disorder, although the severity of it is in the range of moderate to major. This was concurred by Hu *et al.* (2000) who assessed the mental disturbances induced in 41 male construction workers who had witnessed a fatality of a co-worker in a period of one to four months prior to the assessment. The exposed workers had a high rate of post-traumatic stress disorder, which was shown by symptoms such as depressed mood, guilt, insomnia, decreased interest in work and other activities, anxiety, somatisation and gastrointestinal symptoms. Stock *et al.* (2010) claimed that the proportion of workers with post-traumatic stress disorder is high among construction workers and Boschman *et al.* (2013) estimated that symptoms of depression and distress among bricklayers who experienced or witnessed an accident were nearly three to five times higher than those who did not.

Technicians, trade workers and labourers are also placed fairly close to other stress factors at work, which include: workplace harassment and/or bullying, exposure to workplace violence, assault and other stress factors (i.e. dietary or deficiency diseases – bulimia, anorexia). Meliá and Becerril (2007) agreed with this and reported that the style of leadership provided to construction workers, role conflict and bullying behaviour significantly affect workers' experiences of tension and burnout, which in turn affect their psychological health. The ways managers or supervisors treat workers, provide support, resolve problems and/or develop their jobs are important predictors of the level of tension and burnout experienced. Nonetheless, the degree of mental ill-health suffered by them is moderate to major. In terms of age group, 20–29 and 30–39 are mostly represented. Mental health severity level also shows an association with age group, whereby the 20–29 cohort are associated with moderate injuries whilst the other cohort is associated with major injuries.

Combination cloud 4 (Figure 4.7) explains the phenomenon of suicide or attempted suicide among construction workers. Anderson *et al.* (2011) claimed that suicide rates for construction workers occurred at a rate of 40.3 per 100,000 workers per year, which is significantly above the overall national rate for males, at 16.8 per 100,000. Similarly, Doran *et al.* (2015) found that, in 2012, a total of 169 construction workers committed suicide in the Australian construction industry and that for every suicide there were 15 attempts with three (17 per cent) resulting in full incapacity and 12 (83 per cent) resulting in absence from work. The combination cloud reveals that suicide or attempted suicide is prevalent among young workers of under 20 compared with other age groups, and the lowest weekly income is associated with this group. These workers are more likely to be apprentices. Higher rates of suicide in younger construction workers may be due to pressures associated with joining a 'masculine' industry where there is a considerable amount of bullying, especially towards young apprentices (Heller *et al.* 2007). Bullying in the workplace has been proven prevalent in blue-collar working

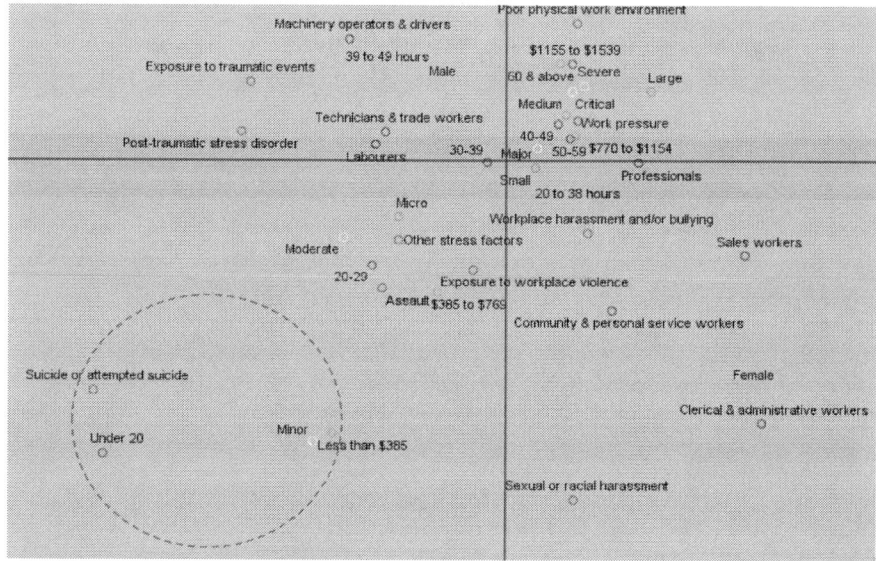

Figure 4.7 Combination cloud 4.

environments such as construction (Agervold and Mikkelsen 2004) and has also been linked to suicide in the UK and Norway (Rayner *et al.* 2002).

Combination cloud 5 (Figure 4.8) indicates that females who are largely in roles such as clerical/administration or community/personal services in the Australian construction industry are likely to experience mental stress due to sexual or racial harassment at work. However, the intensity of psychological symptoms suffered is largely minor to moderate. The obvious reason that could be derived from this cloud is that female employees in the Australian construction industry are underrepresented in mainstream construction and engineering roles that are associated with greater work pressure and long working days. Lingard and Francis (2004) confirmed that site-based women do not experience any greater difficulty with the work–family interface than women based in the regional or head office of a construction organisation. However, Bowen *et al.* (2014a) stated that female professionals report significantly higher levels of stress than male professionals in the construction industry, because in the masculine industry they need to prove their worthiness in the workplace in order to have job security, which is referred to as overcompensating. There is a lack of information regarding female construction professionals and their stress factors, and in-depth studies are needed to uncover this.

Goldenhar *et al.* (1998) investigated psychological symptoms experienced by female construction workers and found sexual harassment, gender discrimination, skill underutilisation and overcompensating to prove themselves to co-worker and supervisors to be the common stress factors encountered.

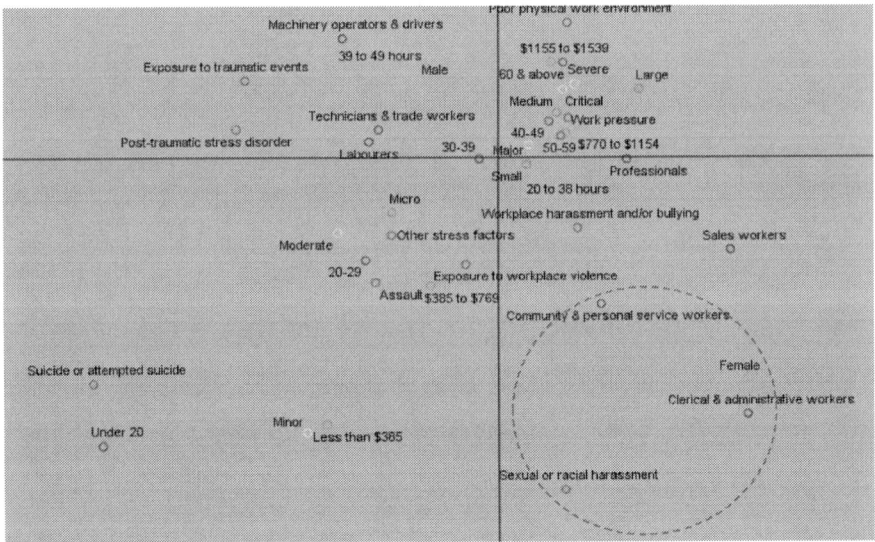

Figure 4.8 Combination cloud 5.

Theoretical expansion

This section compares and contrasts the data mining discoveries with the existing theories of work stress. Many theories of work stress are available in the occupational psychology literature and these are grouped under three broad categories, as shown in Table 4.1, namely: interactional theories, transactional theories and multi-perspective theories. Each theory tries to explain the source of work stress form a particular angle. A critical analysis of these theories in light of the findings discussed in this chapter suggests that among the 15 theories discussed in Table 4.1, Cooper and Marshall's model appears to provide a broader picture of the causes of work stress in construction.

The theory postulates that work stress develops by the interaction of stressors at work, personal life stressors and individual characteristics (see Figure 4.9). Stressors related to work include:

- factors intrinsic to the job (e.g. poor working conditions, time pressures, work overload, physical danger, etc.);
- role in the organisation (e.g. role conflict, role ambiguity, responsibility for people, etc.);
- relationships at work (e.g. poor relations with superiors, subordinates or colleagues, difficulties delegating responsibilities, etc.);
- career development (e.g. under-promotion, lack of job security, thwarted ambitions, etc.); and
- organisational structure and climate (e.g. decision-making attitude and office politics).

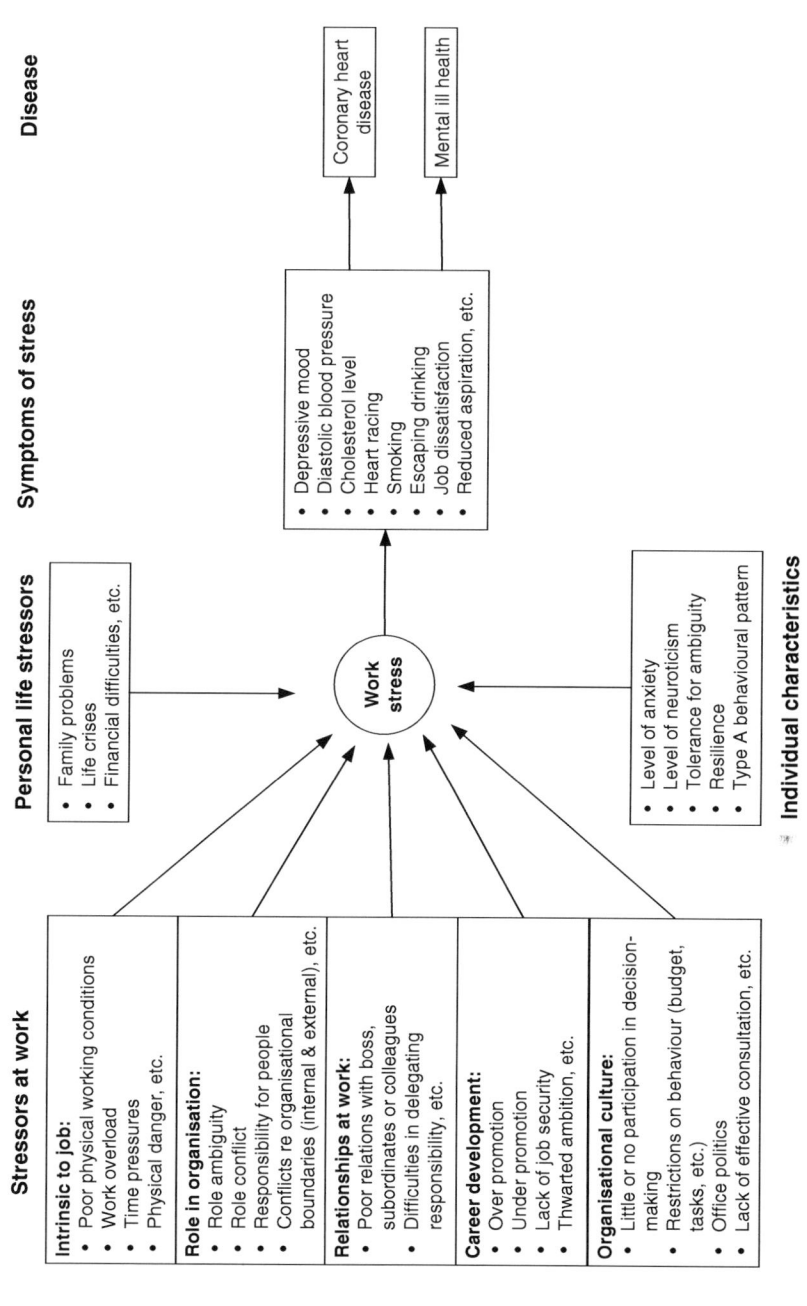

Figure 4.9 Work stress model.

Source: adapted from Cooper and Marshall 1976, p. 12.

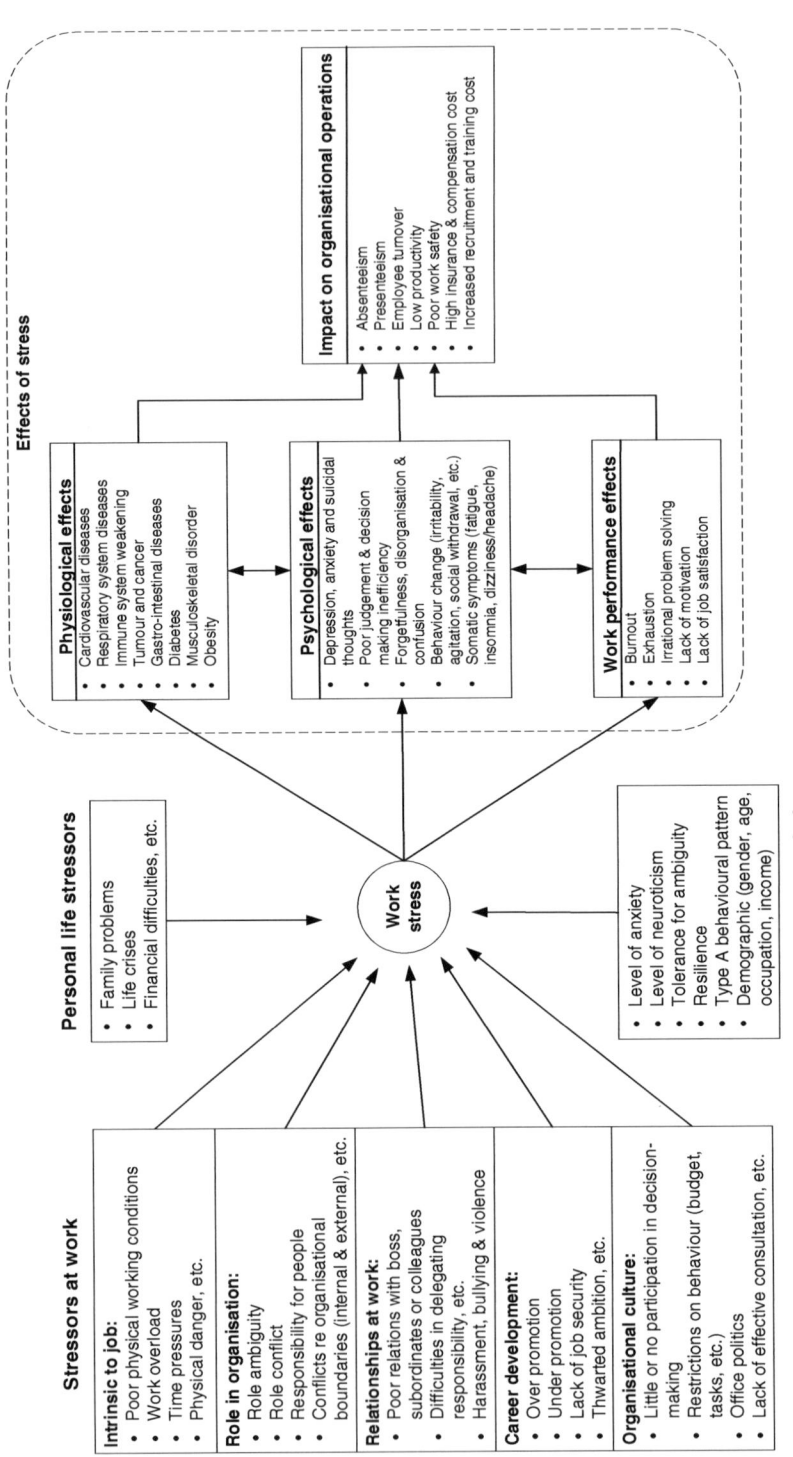

Figure 4.10 Extended work stress model.

Among the personal life stressors are: family problems, life crises, financial difficulties, etc. Work and personal life are two avenues where individuals find life satisfaction. When the work life is stressful, personal life satisfaction balances it. However, when both work and personal life are unfavourable, stress is multiplied. Individual characteristics also play a role in amplifying or subduing the stress. Depending on type of personality (Type A, B or C), and the level of tolerance/resilience, anxiety and neuroticism, stress is coped with or exacerbated. Uncontrolled stress results in symptoms such as depressive mood, diastolic blood pressure, heart racing, smoking, escapist drinking, job dissatisfaction, reduced aspiration, etc. Long-suffered symptoms eventually lead to mental ill-health and coronary heart diseases to employees.

The discoveries of the data mining generally support the stressors intrinsic to work, although the list under the sub-factor of 'relationships at work' can be extended by adding 'harassment, bullying and violence'. No knowledge related to personal life stressors and individual characteristics could be discovered. This is mainly due to the lack of relevant information in the dataset utilised. However, the relationship between demographic factors (gender, age, occupation and income) and work stress as discovered in this study is not represented in Cooper and Marshall's model and it could be a potential addition to it. Moreover, the effects of suffering work stress on individuals are limited in the model. A detailed account of the effects of work stress on individuals and organisations has been discussed in an earlier section of the chapter, and could be absorbed into the model. Hence, Figure 4.10 illustrates an 'extended work stress model'.

Conclusion

Construction is a high-risk industry for work-related fatalities, injuries and mental ill-health. Whilst research on construction fatalities and injuries is prolific, studies on psychological well-being of construction professionals and workers are limited. Data mining of workers' compensation records has revealed valuable insights into the nature and patterns of psychological injuries reported in the Australian construction industry. Mental ill-health is suffered by people at all levels in the construction industry; top managers, middle manager, line managers or supervisors, and operatives. However, the causes and severities vary considerably.

- Work pressure is responsible for most severe mental ill-health in construction and it largely affects construction professionals (middle and line managers) in all sizes of organisations, particularly those who are site-based. Work pressure is exerted by multiple factors, including: high workload, time pressure, working long hours and insufficient time for family/home.
- Top managers in the construction industry suffer largely from adjustment disorders. Strong influences of external factors such as government policies, infrastructure investment decisions, global changes and market competition

on construction businesses and the tendency to keep pace with such changes drive top managers to maintain agility and remain dynamic in their strategies, focus and approaches. This would mean they are often faced with a new learning curve and rapid change, resulting in adjustment disorders. As for site-based managers, dealing with a dynamic mix of subcontractors, workforce, activities and challenges throughout the project duration may cause adjustment disorders. Whilst experienced managers could manage the rapid, continual change effectively, their less experienced counterparts may struggle; the industry features a significant proportion of such managers.

- Primary mental stress factors facing construction operatives are exposure to traumatic events, assault, exposure to workplace violence, workplace harassment and/or bullying. Even though the stressors are multiple, the overall intensity of psychological symptoms suffered is not as severe as that of professionals and managers. Nonetheless, as construction is a labour-intensive, harsh industry, controlling these stressors is critical because a profile of numerous moderately intense cases is as bad as a profile of few severely intense cases.
- Suicidal ideation is predominant among young construction workers who may enter the industry as apprentices and encounter challenges in coping with the masculine and bullying culture of the industry.
- There is a significant under-representation of females in mainstream construction roles in the industry; yet those who largely work in administration and clerical roles develop psychological symptoms due to experiencing sexual harassment and gender-based discrimination. The minority of female construction professionals and operatives experience work stress in over-compensating to prove their worthiness to the workplace.

Safe Work Australia (2013) estimated that loss of productivity due to psychological injuries is costing Australian businesses in excess of $10 billion per year. A significant proportion of workers who suffer from mental health disorders are facing difficulties to return to work (MacDonald *et al.*, 2003). The cost to society caused due to delays in returning to work or inability to fully return to normal functioning is expensive (Mason *et al.* 2006; Michaels *et al.* 2000). Many of those who return to work are unable to remain in employment because of difficulties meeting social and performance demands of the workplace (Krause *et al.* 2001). They also have greater post-discharge health service utilisation than that of the general population. Therefore, they potentially contribute the least to the economy, but cost more in terms of service provision (Aitken *et al.* 2012). Psychological injuries are also associated with an elevated risk of suicide and physical health problems (Wald and Taylor 2009). This study could contribute to potentially reducing the share of the construction industry to this socio-economic burden in the following ways.

- The study has identified that work pressure is the main cause of the costliest psychological injuries in the construction industry. WHS and employee welfare authorities could use this finding to regulate maximum allowable work hours per day on construction sites and introduce 'manageable workload audit schemes' to safeguard the well-being of construction professionals.

- The study has recognised that less experienced managers could benefit from their more experienced counterpart in managing rapid changes and dynamics in construction and thereby mitigate the probability of suffering adjustment disorders. Construction organisations may put this insight into practice by infusing quasi-formal support and mentor schemes.
- Overcompensating is a significant stress factor challenging female construction professionals and operatives. Construction organisations and employee welfare authorities could promote a female-friendly work culture that recognises the differences and sets performance expectations accordingly, which could improve the proportion of women in construction.
- Induction schemes may be set up by construction organisations to inform and educate senior operatives on the acceptable practices of treating apprentices and the socio-economic and legal consequences of inappropriate practices that may push apprentices to suicidal ideation.
- The findings send a strong message that construction sites are full of harassment, violence and bullying. Construction organisations and WHS authorities may introduce systems and mechanisms that militate against the presence of thug culture in the construction industry.

The contribution to knowledge from this research is manifold. First, there is the knowledge discovered concerning work stress in the construction industry; the varying patterns of causes and effects of work stress among different individuals (managers, professionals, supervisors, operatives and female). Second, there is the expansion of Cooper and Marshall's theory and the development of a comprehensive work stress model that has the potential for application in various fields and workplaces, whether it is project-based such as construction or an office-based environment. The theory provides a good basis for conducting investigation of mental stress and its effect on employees and organisations. Finally, the multiple correspondence analysis technique has gained much application in many fields as an analytic tool but, to date, it is an overlooked method in construction broadly, and specifically in the health and safety research domain. The research therefore provides fresh insights for construction health and safety researchers and serves as an exemplary application.

The data mining exercise reported in this chapter has discovered some patterns of psychological injuries in the construction industry. Building on these findings, further research could be undertaken in the following topics:

- Investigations into work stress, its effects and coping strategies among female construction professionals.
- Roles of personal life and individual characteristics in altering work stress among construction professionals and workers.

- Testing the applicability of a proposed extended work stress model
- Development of an online stress audit and management system, supported by a mobile app, that enables construction organisations to identify professionals and managers at risk of psychological ill-health and thereby eliminate the risk proactively.

References

Abramis, D.J. (1994). Relationship of job stressors to job performance: Linear or an inverted-u? *Psychological reports*, 75(1): 547–558.

Agervold, M. and Mikkelsen, E.G. (2004). Relationships between bullying, psychosocial work environment and individual stress reactions. *Work & Stress*, 18: 336–351.

Aitken, L.M., Chaboyer, W., Kendall, E., and Burmeister, E. (2012). Health status after traumatic injury. *Journal of Trauma and Acute Care Surgery*, 72(6): 1702–1708.

Anderson, N.J., Bonauto, D.K., and Adams, D. (2011). Psychiatric diagnoses after hospitalization with work-related burn injuries in Washington State. *Journal of Burn Care & Research*, 32(3): 369–378.

Arquillos, A.L., Romero, J.C.R. and Gibb, A. (2012). Analysis of construction accidents in Spain, 2003–2008. *Journal of Safety Research*, 43: 381–388.

Bakker, A.B. and Demerouti, E. (2007). The job demands-resources model: State of the art. *Journal of Managerial Psychology*, 22(3): 309–328.

Belkić, K., Schnall, P., Landsbergis, P., and Baker, D. (2000). The workplace and cardiovascular health: Conclusions and thoughts for a future agenda. *Occupational Medicine (Philadelphia, PA.)*, 15(1): 307–321, v–vi.

Bongers, P.M., Winter, C.R.d., Kompier, M.A.J., and Hildebrandt, V.H. (1993). Psychosocial factors at work and musculoskeletal disease. *Scandinavian Journal of Work, Environment & Health*, 19(5): 297–312.

Boschman, J.S., van der Molen, H.F., Sluiter, J.K., and Frings-Dresen, M.H.W. (2013). Psychosocial work environment and mental health among construction workers. *Applied Ergonomics*, 44(5): 748–755.

Bowen, P., Edwards, P., Lingard, H., and Cattell, K. (2014a). Occupational stress and job demand, control and support factors among construction project consultants. *International Journal of Project Management*, 32(7): 1273–1284.

Bowen, P., Govender, R., and Edwaards, P. (2014b). Structural equation modelling of occupational stress in the construction industry. *Journal of Construction Engineering and Management*, 140(9): 1–14.

Bright, J. (2001). Individual difference factors and stress; A case study paper. National Occupational Health and Safety Commission Symposium on the OSH Implications of Stress. December 2001. Melbourne: NOHSC.

Carayon, P., Smith, M.J., and Haims, M.C. (1999). Work organization, job stress, and work-related musculoskeletal disorders. *Human Factors: The Journal of the Human Factors and Ergonomics Society*, 41(4): 644–663.

Chen, W-Q., Wong, T-W., and Yu, T-S. (2009). Direct and interactive effects of occupational stress and coping on ulcer-like symptoms among Chinese male off-shore oil workers. *American Journal of Industrial Medicine*, 52(6): 500–508.

Cohen, S. and Williamson, G.M. (1991). Stress and infectious disease in humans. *Psychological Bulletin*, 109(1): 5.

Cooper, C.L., Liukkonen, P., and Cartwright, S. (1996). *Stress Prevention in the Work-place: Assessing the Costs and Benefits to Organisations*. Dublin: European Foundation for the Improvement of Living and Working Conditions.

Cooper, C.L., and Marshall, J. (1976). Occupational sources of stress: A review of the literature relating to coronary heart disease and mental ill health. In C.L. Cooper (Ed.), *From Stress to Wellbeing* (Vol. 1). UK: Palgrave Macmillan.

Costa, P.S., Santos, N.C., Cunha, P., Cotter, J., and Sousa, N. (2013). The use of multiple correspondence analysis to explore associations between categories of qualitative variables in healthy ageing. *Journal of Aging Research*, 2013, Article ID 302163, pp. 1–12.

Cox, T. (1993). *Stress Research and Stress Management: Putting Theory to Work*. Sudbury: HSE Books.

Cox, T., and Griffiths, A. (2010). Work-related stress: A theoretical perspective. In S. Leka and J. Houdmont (Eds), *Occupational Health Psychology* (pp. 31–56). UK: Wiley-Blackwell.

Cox, T., Cox, S., and Thirlaway, M. (1983). The psychological and physiological response to stress. *Physiological Correlates of Human Behaviour*. London: Academic Press.

Cox, T., Griffiths, A., and Rial-González, E. (2000). Research on Work-related Stress: European Agency for Safety and Health at Work. Belgium: European Agency for Safety and Health at Work.

Cox, T., Griffiths, A., and Leka, S. (2005). Work organization and work-related stress. In K. Gardiner and J.M. Harrington (Eds), *Occupational Hygiene* (3rd edn, pp. 421–432). Oxford, UK: Blackwell.

Das, S. and Sun, X. (2015). Association knowledge for fatal run-off-road crashes by multiple correspondence analysis. IATSS Research, http://dx.doi.org/10.1016/j.iatssr.2015.07.001. pp. 1–10.

de Jonge, J., and Dormann, C. (2003). The DISC model: Demand-induced strain compensation mechanisms in job stress. In M.F. Dollard, A.H. Winefield and H.R. Winefield (Eds), *Occupational Stress in the Service Professions* (pp. 43–74). London: Taylor & Francis.

Demerouti, E., Bakker, A.B., Nachreiner, F., and Schaufeli, W.B. (2001). The job demands-resources model of burnout. *Journal of Applied Psychology*, 86(3), pp. 499.

Dewe, P., Cox, T., and Leiter, M. (2000). *Coping, Health and Organizations. Issues in Occupational Health*, London: Taylor & Francis.

Dewe, P.J., O'Driscoll, M.P., and Cooper, C.L. (2012). Theories of psychological stress at work. In R.J. Gatchel and I.Z. Schultz (Eds), *Handbook of Occupational Health and Wellness* (pp. 23–38). New York: Springer.

Dimsdale, J.E. (2008). Psychological stress and cardiovascular disease. *Journal of the American College of Cardiology*, 51(13): 1237–1246.

Dollard, M. (2001). Work stress theory and intervention: From evidence to policy a case study. In The National Occupational Health and Safety Commission Symposium on the OSH Implications of Stress. December 2001. Melbourne: NOHSC.

Doran, C.M, Ling, R., and Milner, A. (2015). The economic cost of suicide and non-fatal behaviour in the Australian construction industry by state and territory. www.matesinconstruction.org.au/flux-content/mic-2013/pdf/Cost-of-suicide-in-construction-industry-final-report.pdf (accessed 20 February 2016).

Dumrak, J., Mostafa, S., Kamardeen, I., and Rameezdeen, R. (2013). Factors associated with the severity of construction accidents: The case of South Australia. *Construction Economics and Building*, 13(4): 32–49.

Edwards, J.R. (1992). A cybernetic theory of stress, coping, and well-being in organizations. *The Academy of Management Review*, 17(2): 238–274.

Eysenck, M.W., Derakshan, N., Santos, R., and Calvo, M.G. (2007). Anxiety and cognitive performance: Attentional control theory. *Emotion*, 7(2): 336.

Flier, J.S., Underhill, L.H., and McEwen, B.S. (1998). Protective and damaging effects of stress mediators. *New England Journal of Medicine*, 338(3): 171–179.

French, J.R.P., Jr., Caplan, R.D., and Harrison, R.V. (1982). *The Mechanisms of Job Stress and Strain*. London: Wiley.

Fung, I.W.H., Tam, V.W.Y., Lo, T.Y., and Lu, L.L.H. (2009). Developing a Risk Assessment Model for construction safety. *International Journal of Project Management*, 28(6): 593–600.

Gangolells, M., Casals, M., Forcada, N., Roca, X., and Fuertes, A. (2010). Mitigating construction safety risks using prevention through design. *Journal of Safety Research*, 41(2): 107–122.

Ganster, D.C. and Rosen, C.C. (2013). Work stress and employee health: A multidisciplinary review. *Journal of Management*, 39(5): 1085–1117.

Glaser, R., Padgett, D.A., Litsky, M.L., Baiocchi, R.A., Yang, E.V., Chen, M., Yeh, P-E., Klimas, N.G., Marshall, G.D., and Whiteside, T. (2005). Stress-associated changes in the steady-state expression of latent Epstein–Barr virus: Implications for chronic fatigue syndrome and cancer. *Brain, Behavior, and Immunity*, 19(2): 91–103.

Godbout, J.P. and Glaser, R. (2006). Stress-induced immune dysregulation: Implications for wound healing, infectious disease and cancer. *Journal of Neuroimmune Pharmacology*, 1(4): 421–427.

Goldenhar, L.M., Swanson, N.G., Hurrell Jr, J.J., Ruder, A., and Deddens, J. (1998). Stressors and adverse outcomes for female construction workers. *Journal of Occupational Health Psychology*, 3(1): 19–32.

Greenacre, M. and Blasius, J. (2006). *Multiple Correspondence Analysis and Related Methods*. London: Chapman & Hall/CRC.

Hart, P.M. and Cooper, C.L. (2001). Occupational stress: Toward a more integrated framework. *Handbook of Industrial, Work and Organizational Psychology*, 2: 93–114.

Haynes, N.S., and Love, P.E.D. (2004). Psychological adjustment and coping among construction project managers. *Construction Management and Economics*, 22(2): 129–140.

Headey, B. (2006). Subjective well-being: Revisions to dynamic equilibrium theory using national panel data and panel regression methods. *Social Indicators Research*, 79(3): 369–403.

Headey, B., and Wearing, A. (1989). Personality, life events, and subjective well-being: Toward a dynamic equilibrium model. *Journal of Personality and Social Psychology*, 57(4): 731.

Health and Safety Executive (2001). *Work-Related Stress: A Short Guide*. Sudbury: HSE Books.

Heller, T.S., Hawgood, J.L., and Leo. D.D. (2007). Correlates of suicide in building industry workers. *Archives of Suicide Research*, 11(1): 105–117.

Hu, B.S., Liang, Y.X, Hu, X.Y., Long, Y.F., and Ge, L.N. (2000). Posttraumatic stress disorder in co-workers following exposure to a fatal construction accident in China. *International Journal of Occupational and Environmental Health*, 6(3): 203–207.

IBM (n.d.). Help – IBM SPSS Statistics URL: http://127.0.0.1:56640/help/index.jsp?topic=%2Fcom. ibm.spss.statistics.cs%2Fspss%2Ftutorials%2Fcasestudies_intro.htm (accessed 20 October 2016).

Jacobs, M.A., Spilken, A.Z., Norman, M.M., and Anderson, L.S. (1970). Life stress and respiratory illness. *Psychosomatic Medicine*, 32(3): 233–242.

Jamal, M. (1984). Job stress and job performance controversy: An empirical assessment. *Organizational Behavior and Human Performance*, 33(1): 1–21.

Jex, S.M (1998). *Stress and Job Performance: Theory, Research, and Implications for Managerial Practice*. Thousand Oaks, CA: Sage Publications.

Kalantaridou, S.N., Makrigiannakis, A., Zoumakis, E., and Chrousos, G.P. (2004). Stress and the female reproductive system. *Journal of Reproductive Immunology*, 62(1–2): 61–8.

Kamardeen, I., and Rameezdeen, R. (2016). Causation model for psychological injuries in the construction industry. In *Proceedings of the CIB World Building Congress 2016*, Vol. 5 (pp. 616–628). Finland, CIB.

Karasek, R.A. and Theorell, T. (1990). *Healthy Work: Stress, Productivity and the Reconstruction of Working Life*. New York: Basic Books.

Kawakami, N. and Haratani, T. (1999). Epidemiology of job stress and health in Japan: Review of current evidence and future direction. *Industrial Health*, 37(2): 174–186.

Keegel, T., Ostry, A., and LaMontagne, A.D. (2009). Job strain exposures vs. Stress-related workers' compensation claims in Victoria, Australia: Developing a public health response to job stress. *Journal of Public Health Policy*, 30(1): 17–39.

Krause, N., Frank, J., Dasinger, L., Sullivan, T.J., and Sinclair, S.J. (2001). Determinants of duration of disability and return-to-work after work-related injury and illness: Challenges for future research. *American Journal of Industrial Medicine*, 40(4): 464–484.

Lazarus, R.S., and Folkman, S. (1984). *Stress, Appraisal, and Coping*. New York: Springer.

Leung, M., Chan, I.Y.S., and Cooper, C.L. (2015). *Stress Management in the Construction Industry*. West Sussex: Wiley–Blackwell.

Leung, M.Y., Chan, Y.S., and Yuen, K.W. (2010). Impacts of stressors and stress on the injury incidents of construction workers in Hong Kong. *Journal of Construction Engineering and Management*, 136: 1093–1103.

Leung, M.-Y., Chan, Y.-S., and Olomolaiye, P. (2008). Impact of stress on the performance of construction project managers. *Journal of Construction Engineering and Management*, 134(8): 644–652.

Lindquist, T.L., Beilin, L.J., and Knuiman, M.W. (1997). Influence of lifestyle, coping, and job stress on blood pressure in men and women. *Hypertension*, 29(1): 1–7.

Lingard, H. and Francis, V. (2004). The work-life experiences of office and site-based employees in the Australian construction industry. *Construction Management and Economics*, 22(9): 991–1002.

Lupien, S.J., McEwen, B.S, Gunnar, M.R., and Heim, C. (2009). Effects of stress throughout the lifespan on the brain, behaviour and cognition. *National Review of Neuroscience*, 10(6): 434–445.

MacDonald, H.A., Colotla, V., Flamer, S., and Karlinsky, H. (2003). Posttraumatic stress disorder (PTSD) in the workplace: A descriptive study of workers experiencing PTSD resulting from work injury. *Journal of Occupational Rehabilitation*, 13(2): 63–77.

Macklin, D.S., Smith, L.A. and Dollard, M.F (2006) Public and private sector work stress: Workers compensation, levels of distress and job satisfaction, and the demand-control-support model. *Australian Journal of Psychology*, 58(3): 130–143.

Manning, M.R., Jackson, C.N., and Fusilier, M.R. (1996). Occupational stress, social support, and the costs of health care. *Academy of Management Journal*, 39(3): 738–750.

Mark, G.M., and Smith, A.P. (2008). Stress models: A review and suggested new direction. *Occupational Health Psychology*, 3: 111–144.

Marshall, V., Heinz, W.R., Kruger, H., and Verma, A. (2001). *Restructuring Work and the Life Course*. Toronto, ON: University of Toronto Press.

Maslach, C. (1998). A multidimensional theory of burnout, In C.L. Cooper (Ed.), *Theories of Organizational Stress* (pp. 68–85). New York: Oxford University Press.

Mason, S., Turpin, G., Woods, D., Wardrope, J., and Rowlands, A. (2006). Risk factors for psychological distress following injury. *British Journal of Clinical Psychology*, 45(Pt 2): 217–230.

McEwen, B.S. (2006). Sleep deprivation as a neurobiologic and physiologic stressor: Allostasis and allostatic load. *Metabolism*, 55(supplement 2): S20–S23.

McEwen, B.S. (2007). Physiology and neurobiology of stress and adaptation: Central role of the brain. *Physiological Reviews*, 87(3): 873–904.

McEwen, B.S. (2008). Central effects of stress hormones in health and disease: Understanding the protective and damaging effects of stress and stress mediators. *European Journal of Pharmacology*, 583(2–3): 174–185.

Melchior, M., Caspi, A., Milne, B.J., Danese, A., Poulton, R., and Moffitt, T.E. (2007). Work stress precipitates depression and anxiety in young, working women and men. *Psychological Medicine*, 37(8): 1119–1129.

Meliá, J.L and Becerril, M. (2007). Psychosocial sources of stress and burnout in the construction sector: A structural equation model. *Psicothema*, 9(4): 679–686.

Michaels, A.J., Michaels, C.E., Smith, J.S., Moon, C.H., Peterson, C., and Long, W.B. (2000). Outcome from injury: General health, work status, and satisfaction 12 months after trauma. *The Journal of Trauma: Injury, Infection, and Critical Care*, 48(5): 841–850.

Miller, T.A., and McCool, S.F. (2003). Coping with stress in outdoor recreational settings: An application of transactional stress theory. *Leisure Sciences*, 25(2–3): 257–275.

National Institute of Occupational Health and Safety (NIOSH) (1999). *Stress at Work*. Cincinnati: NIOSH.

National Patient Safety Agency (2008). A risk matrix for risk managers. www.nrls.npsa. nhs.uk/EasySiteWeb/getresource.axd?AssetID=60149&... (accessed 10 April 2014).

Nomura, K., Nakao, M., Sato, M., Ishikawa, H., and Yano, E. (2007). The association of the reporting of somatic symptoms with job stress and active coping among Japanese white-collar workers. *Journal of Occupational Health*, 49(5): 370–375.

Rada, R.E. and Johnson-Leong, C. (2004). Stress, burnout, anxiety and depression among dentists. *The Journal of the American Dental Association*, 135(6): 788–794.

Räihä, I., Kemppainen, H., Kaprio, J., Koskenvuo, M., and Sourander, L. (1998). Lifestyle, stress, and genes in peptic ulcer disease: A nationwide twin cohort study. *Archives of Internal Medicine*, 158(7): 698–704.

Rayner, C., Howl, H., and Cooper, C. L. (2002). *Work-place Bullying: What We Know, Who is to Blame, and What Can We Do?* London: Taylor and Francis.

Safe Work Australia (2013). Media release: Mental stress costs Australian businesses more than $10 billion per year. www.safeworkaustralia.gov.au/sites/SWA/media-events/media-releases/Documents/2013%20Media%20Releases/MR08042013-Mental-Stress-Cost-Australian-Businesses.pdf (accessed 17 May 2015).

Schabracq, M.J., Winnubst, J.A., and Cooper, C.L. (2003). *The Handbook of Work and Health Psychology*, UK: John Wiley & Sons.

Siegrist, J. (2009). Job control and reward: effects on well being. In S. Cartwright and C.L. Cooper (Eds), *The Oxford Handbook of Organizational Well-being* (pp. 109–132). Oxford: Oxford University Press.

Siegrist, J. (1996). Adverse health effects of high-effort/low reward conditions. *Journal of Occupational Health Psychology*, 1: 27–41.

Spector, P.E., and Jex, S.M. (1998). Development of four self-report measures of job stressors and strain: Interpersonal Conflict at Work Scale, Organizational Constraints Scale, Quantitative Workload Inventory, and Physical Symptoms Inventory. *Journal of Occupational Health Psychology*, 3(4): 356–367.

Stanghellini, V. (1998). Relationship between upper gastrointestinal symptoms and lifestyle, psychosocial factors and comorbidity in the general population: Results from the Domestic/International Gastroenterology Surveillance Study (DIGEST). *Scandinavian Journal of Gastroenterology. Supplement*, 231: 29–37.

Steptoe, A. and Kivimäki, M. (2012). Stress and cardiovascular disease. *Nature Reviews Cardiology*, 9(6): 360–370.

Stock, S.J., McNamee, R., Carder, M., and Agius, R.M. (2010). The incidence of medically reported work-related ill health in the UK construction industry. *Occupational and Environmental Medicine*, 68: 547–576.

Sutherland, V. and Davidson, M.J. (1993). Using a stress audit: the construction site manager experience in the UK. *Construction Management and Economics*, 22(3): 273–286.

Wald, J. and Taylor, S. (2009). Work impairment and disability in posttraumatic stress disorder: A review and recommendations for psychological injury research and practice. *Psychological Injury and Law*, 2(3): 254–262.

Warr, P. (1987). *Work, Unemployment, and Mental Health*. Oxford: Oxford University Press.

WorkCover NSW (2014). Overview of work-related stress. www.workcover.nsw.gov.au/__data/assets/pdf_file/0007/19915/tip-sheet-1-work-related-stress-10731.pdf (accessed 29 May 2015).

5 Predicting and preventing secondary psychological injuries in construction using analytics

Introduction

Existing scientific evidence supports a bi-directional relationship between psychological disorders and work-related injuries, i.e. pre-existing psychological disorders lead to a high rate of work-related injuries; likewise suffering work-related injuries elevates the likelihood of developing post-injury psychological disorders. A non-construction US study, for example, found that injured workers were 44 per cent more likely to be treated for outpatient depression than the non-injured workers at three-month follow-up (Asfaw and Souza 2012). Earlier research by MacDonald *et al.* (2003) found that 55 per cent of workers were diagnosed with post-traumatic stress disorder after a work-related injury and as such faced difficulties returning to work. Many of those who returned to work were unable to remain in employment because of difficulties meeting the social and performance demands of the workplace (Krause *et al.* 2001). They also had greater post-discharge health service utilisation than that of the general population. Therefore, they potentially contribute at a lower level to the economy but cost more in terms of service provision (Aitken *et al.* 2012).

Since the construction sector records one of the highest workplace incident rates, studying the after-effects of injuries is crucial, i.e. the link between workplace injuries and the onset of psychological disorders. Nevertheless, hardly any studies have been undertaken to date to investigate the issue in construction. Previous research on health and safety in the construction industry focused on: (1) physical injuries and health; (2) psychological injuries caused by work stressors; and (3) relationships between mental stress and accidents. Research is warranted to understand the issue of post-injury psychological disorders in the construction industry context. Hence, this chapter aims to address the following research questions.

- Why are operatives affected by psychological disorders post work injury?
- What method can be applied to pro-actively identify operatives at risk of post-injury psychological disorders?
- What early preventive methods might be implemented?

Secondary psychological injuries

There are two types of psychological injuries suffered by workers: primary and secondary. Primary psychological injuries refer to psychological disorders, such as acute stress reaction, anxiety, burnout, trauma and depression, that are caused directly by workplace stressors, such as work pressure, poor work environment and conditions, harassment, bullying, exposure to traumatic events or violence at work, and lack of job support (WorkCover South Australia 2012). The previous chapter discussed in detail the causes and effects of primary psychological injuries.

Psychological disorders that develop due to suffering physical injuries or illnesses and challenges resulting from the injuries/illnesses, such as changes to socio-economic and work conditions and quality of life, are known as Secondary Psychological Injuries (SPIs), which develop over a long period following work-related physical injuries or illnesses (WorkCover Queensland 2015). SPIs pose long-term harmful consequences for workers, their families and employers. Nearly half of injured workers suffer subsequent psychological disorders (Asfaw and Souza 2012; MacDonald *et al.* 2003). WorkCover Queensland (2015) estimated that, on average, workers who reported an SPI were absent from work twice the period of those who suffered a physical work injury only. Death from suicide was found to be up to five times higher among SPI sufferers than the general population and it is most likely to occur within 5–6 years post injury (Dezarnaulds and Ilchef 2014).

SPI is a broad term that refers to any form of mental health issue suffered after a workplace injury or illness, and the commonly reported disorders are: anxiety, depression and post-traumatic stress disorder (Kendrick *et al.* 2011). The following sections explain these psychiatric disorders with their symptoms.

Anxiety

Anxious feelings of stress and worry are normal reactions to stressful situations where one feels under pressure, and they disappear once the stressor has passed. However, employees with anxiety disorder experience the anxious feelings for a long period after the stressful event for no apparent reasons. Anxiety disorder can manifest itself in different forms, such as uncontrollable worry, phobia (irrational intense fear), panic attacks, obsessive-compulsive disorder and nightmares or flashbacks of a traumatic event. The following common symptoms are experienced by affected people in a constant fashion (Beyondblue n.d.; American Psychiatric Association 1994):

- Restlessness or feeling on edge/irritable
- Racing heart
- Tightening of the chest
- Fatigue
- Hot and cold flushes

- Difficulties with concentration
- Excessive muscle tension
- Disturbed sleep
- Persistent irrational thought or repetitive behaviours that are excessive/irrational

In workplace settings, all employees feel occasionally anxious and an optimum level of anxiety is beneficial as it increases the level of arousal, helps them perform at peak efficiency and improves performance. However, beyond the optimum point, it is counterproductive and psychologically damaging (Sullivan 2004). Among the factors that cause anxiety disorder at work are: excessive work demand (workload, the pace of work and deadlines), perceived lack of control over work/life, dangerous lines of work and job insecurity (Braverman 1992). Injured workers who sustained a disability, impairment or chronic pain could face most of the factors above.

Post-traumatic stress disorder

Post-Traumatic Stress Disorder (PTSD) is characterised by the development of long lasting anxiety response by an individual, who has been exposed to a traumatic event in the following manner (APA 1994; Braverman 1992):

- The person experienced first-hand or witnessed an event that involved actual or threatened death or serious injury, or a threat to the physical integrity of self or others; and
- The person's response involved intense fear, helplessness and horror.

Symptoms of PTSD can be of the following different forms (Beyondblue n.d.; Braverman 1992):

- Re-experiencing the traumatic event and distress persistently – recurrent intrusive memories of the event, including images, thoughts, flashbacks or nightmares; and feeling as if the event were recurring.
- Persistent avoidance of things, activities, places, people and conversations that are associated with or reminders of the traumatic event.
- Persistent symptoms of increased arousal – difficulty sleeping; irritability or outburst of anger; concentration and memory difficulties; and being easily startled.
- Emotional numbness and lack of responsiveness – reflected by a loss of interest in important/previously enjoyable activities; diminished affection (e.g. loss of love feelings); and sense of foreshortened future (e.g. lack of interest in having a career, marriage, children or a normal life span).
- Feelings of detachment from friends and family members (social withdrawal).

It is common to experience the above symptoms immediately after a work-related traumatic incident, but they wear off over time. The continuation of the symptoms beyond one month may suggest that the injured worker is suffering from PTSD. Moreover, studies show that major depression was one of the common psychiatric disorders that occur alongside PTSD (Shih *et al.* 2010; O'Donnell *et al.* 2004).

Depressive disorder

Depressive disorder is an illness of mood and is characterised by persistently intense lowered mood (feeling sad, down or miserable) for no apparent reason, and a loss of interest in work, hobbies and other activities one normally enjoys (Dezarnaulds and Ilchef 2014; Sullivan 2004). While most people experience sadness or a depressive mood from time to time due to confronting unpleasant circumstances, they recover from it within days; individuals with depressive disorder experience these feelings intensely for a long period (weeks, months or even years). A person may be considered depressed if he/she experiences any of the following symptoms for more than two weeks (Dezarnaulds and Ilchef 2014; Beyondblue n.d.; Braverman 1992):

- Feeling unhappy, sad, down or miserable most of the time
- Loss of interest in enjoyable activities
- Withdrawing from close family and friends
- Unable to concentrate and not getting things done at work
- Feeling overwhelmed, indecisive and lacking confidence
- Feeling worthless, helpless and guilty
- Thoughts such as, 'I'm a failure', 'life's not worth living', 'people would be better off without me'
- Bleak and pessimistic views of the future
- Recurrent suicidal thoughts or acts of self-harm
- Alcohol or drug dependence
- Increased irritability, frustration and moodiness
- Insomnia or hypersomnia
- Fatigue or loss of energy
- Disturbed appetite (increased/decreased)
- Significant, rapid weight gain or loss
- Decreased libido

Model of SPI

The primary cause of an SPI is a work-related injury or illness. However, only about 50 per cent of injured workers develop SPIs, depending on the circumstances they encounter post injury or illness. The model illustrated in Figure 5.1 maps out the path from work injury or illness to SPI, encapsulating sequelae and stressors, risk factors and resilience factors.

The model postulates that serious, non-fatal work-related injuries and illnesses can have substantial negative implications for workers in physical, social,

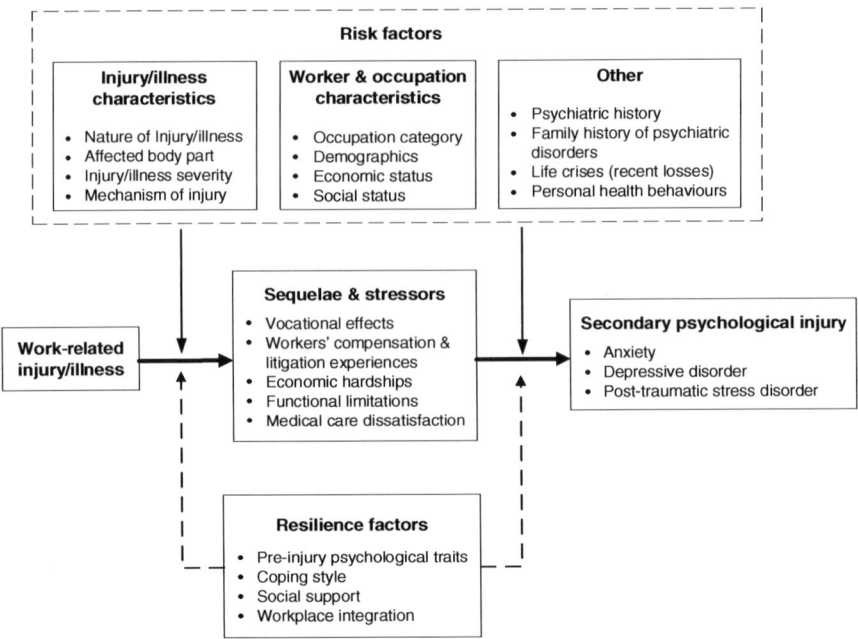

Figure 5.1 Model of SPI.

personal, vocational and economic dimensions, resulting in mental stress and compromised quality of life. Enduring the lowered quality of life with mental stress may lead to developing psychological disorders. The level of changes occurring to the quality of life of suffering workers is influenced by the characteristics of the injury/illness, the workers themselves, the occupation type and pre-existing conditions; referred to as risk factors. On the other hand, psychological and behavioural traits of and external support provided to the injured workers can moderate the negative changes and their psychological outcomes. The sections below explain in detail the negative sequelae of work injuries or illnesses, risk factors and resilience factors.

Sequelae of work-related injuries or illnesses

Workers may suffer permanent or short-term impairments, disabilities and/or chronic pain, as a result of sustaining a work-related injury or illness. These in turn impact on their workers' normal vocational and social functioning as well as socio-economic status. Enduring the stress of unexpected, unpleasant life changes leads to psychological disorders. Previous research in occupational psychology reported numerous ways that the quality of life of such workers is affected and thereby mental stress mounts up.

Vocational effects

Workers with a disability, impairment and/or chronic pain that resulted from work injuries or illnesses may suffer significant mental stress in dealing with disruptions to their work lives, such as: loss of job or job insecurity, change of job role, productivity loss, wage loss, and diminished job satisfaction and motivation.

Severely disabled workers are less likely to return to work, or their employers may not take them back, particularly in fields such as construction where physical ability and independent manoeuvring are essential. Those workers who manage to return to work with a partial disability, impairment or chronic pain and remain in employment have to change the job role and/or reduce their productivity rate and pace. Morse *et al.* (1998) observed that more than one-third of returning workers faced such changes. These changes and work limitations impact on workers' motivation and job satisfaction and even cause wage losses. Reville (1999) confirmed that workers suffer a continued net wage loss for several years after a disabling work injury, with resultant financial consequences. In the long run, they may also face labour market challenges/competition and job insecurity.

In some circumstances, injured workers experience mistrust and stigmatisation at work; for example Tarasuk and Eakin (1995) noted that workers with work-related lower back pain were subject to suspicion by co-workers and supervisors about the legitimacy of their injuries and felt stigmatised. This may lead to losing a job or deciding to change the job or workplace. Those workers who decide to change the job type or industry may face other challenges such as the availability of suitable jobs, lacking skills and experiences for alternative jobs, and ambient economic and labour market conditions (Dembe 2000).

Workers' compensation and litigation experiences

Employers are required by law to procure Workers' Compensation (WC) insurance to protect workers from the financial burden that results from a work-related injury or illness. The insurance typically covers medical expenses and lost wages due to work injury or illness. However, significant obstacles confront injured workers trying to gain what they consider to be rightful benefits through the WC system.

Keogh *et al.* (2000) commented that although the employer's WC insurance is supposed to pay for all medical expenses related to a work-injury or illness, workers had to incur substantial out-of-pocket expenses and/or utilise other private insurance to pay for medical care and treatment. Likewise, WC insurance is supposed to provide lost wages due to injury or illness. However, most WC policies cap the maximum amount of lost wages payable and/or limit the duration, leaving disabled or impaired workers in a net loss of earnings (Reville 1999). Moreover, investigations conducted in the US discovered that workers had to wait up to 20 months to receive wage replacement benefits (wage loss compensation) due to delays in approval and processing (Dawson 1994). Both of these situations leave injured workers in economic hardship and distress. Dembe (2001) reported that a significant proportion of injured workers had problems paying bills and were forced to borrow money as a consequence of their work injuries.

Injured workers must prove occupational causation to obtain benefits from WC insurance, but they often face the unwillingness of employers and WC insurers to accept responsibility for the injury. Moreover, current WC policies do not adequately cover mental health consequences in the rehabilitation of injured workers, leaving workers to deal with these on their own. Hence, delays in treatment and legal disputes are common experiences for injured workers. Involvements in litigation can become a risk factor itself as the legal process requires repeated detailed recall of events and tends to polarise concepts of blame. This process can have a significant impact on the recovery process too (Bay and Donders 2008; Mason *et al.* 2006). Kim (2013) pointed out that additional distress from dealing with WC and litigation play a unique role in SPI onset.

Economic hardships

Workers with a disability or impairment that resulted from a work injury or illness may suffer significant economic hardships, compromised quality of life for the family and thereby mental stress, owing to reduced family income resulting from losing the job, and wage losses and additional expenses for medical care and litigation and other additional expenses related to WC and extra care. Keogh *et al.* (2000) reported that owing to the economic hardship created by job injuries, workers lost homes or had to move their primary residence (downgrade), lost their car and lost health insurance, significantly impacting on the quality of life of the entire family, including children's education and future. These can cause continuous, acute mental stress in workers and their families.

Functional limitations

Workers with a disability, impairment and/or chronic pain that resulted from work injuries or illnesses reported considerable lifelong difficulties in performing daily activities such as household chores, fulfilling other family responsibilities, and recreational and social activities.

Keogh *et al.* (2000) studied workers with upper-extremity cumulative disorders at an average 28 months post their initial workers' compensation claim date and found that over 40 per cent of them were still experiencing hardships with writing with a pen, lifting/carrying a child, cleaning house and placing items onto a high shelf. Likewise, Strunin and Boden (1997) discovered through ethnographic studies of workers with disabling back pain that one-third of them experienced difficulties performing household chores such as cleaning, shopping, removing rubbish and doing laundry. Furthermore, they could not engage in recreational activities such as cycling, bowling, playing volleyball, fishing, camping, and participating in children's activities and events. Added to the distress were difficulties in having sex, and sleeping and getting out of bed in the morning. A separate study by Morse *et al.* (1998) revealed that workers suffering from occupational cumulative disorders have elevated levels of stress at home and are more likely to divorce, and workers with chronic lower back pain have greater levels of family conflicts, while occupational hearing loss weakened the quality of the affected workers' family relationships.

Medical care dissatisfaction

Dembe (1998) reported that patient satisfaction with medical care provided through WC insurance is generally lower than the general healthcare provided for non-occupational conditions. Most injured workers complained that primary treating physicians often did not take their problem seriously, failed to understand the nature of their jobs and did not offer any advice about the prevention of further injury or worsening the current injury. As a result, many of those who returned to work continued to suffer residual pain (Pransky *et al.* 2000). Some workers even face challenges accessing appropriate medical care due to delays of WC providers. In many countries, immigrant workers may face particularly significant barriers to obtaining appropriate medical care for work-related conditions due to restrictions placed on their visas regarding access to public health and other welfare support.

Risk factors

Several factors directly or indirectly moderate the implications and the compromised quality of life suffered by workers. These are called risk factors and are grouped under three broad categories: injury/illness characteristics, worker and occupation characteristics, and other factors.

Injury/illness-related risk factors for SPIs include the nature, cause and severity of the injury/illness and the body part damaged (Sareen *et al.* 2013). Injuries that severely affect the physical appearance and cause loss of consciousness significantly increased the likelihood of psychological symptoms (Lin *et al.* 2014). Synchronously, Sareen *et al.* (2013) stated that burns on hands and face have been linked to PTSD and depression among women. Similarly, traumatic brain injury is specifically linked to PTSD and depression. Workers with upper limb amputations following work injuries reported more symptoms of PTSD and depression than workers with lower limb amputations (Cheung *et al.* 2003). O'Hagan *et al.* (2012) summarised that injuries that result in permanent impairments elevate the risk of major depressive disorder. Likewise, admission to an intensive care unit due to work injury has been shown to be a risk factor for PTSD (O'Donnell *et al.* 2010). Additionally, the level of physical disability and an inability to return to work are associated with an increased risk of depression and anxiety (Esselman *et al.* 2007). In terms of relationships between injury mechanism and SPI, Kellezi *et al.* (2016) noted a higher prevalence of SPIs among fall-related injuries than that of injuries caused by being struck by objects, vehicular incidents and other mechanisms.

Worker and occupation characteristics, namely demographics, occupation type and socio-economic status have indirect moderating effects on the sequelae and SPIs experienced by workers. Several studies have found that demographic factors such as female gender, age, divorced/separated/widowed marital status and low income are significantly associated with psychiatric disorders after a traumatic work-related injury, while low education had a moderate association (Shih *et al.* 2010; Kuo *et al.* 2007; Chang *et al.* 2005; Yang *et al.* 2003). Injured workers with a low family income or who are economically disadvantaged pose a higher risk of depression (Kim 2013). As for the occupation-related risk factors, Kim (2013) found a strong association between post-injury depression, and part-time/seasonal work, shorter job tenure and long working hours. Surprisingly, Kim further claimed that workers in the white-collar and service occupations had higher odds of depression compared with those in blue-collar occupations. This is counter-intuitive as low income is a risk factor for SPIs and is often associated with blue-collar employment. In a reconciling way, O'Hagan *et al.* (2012) stated that there is no relationship between education or occupational class and post-injury mental health issues.

As for the other factors, Cameron *et al.* (2006) argued that pre-existing mental health conditions are a co-founder of the relationship between an injury and poor mental health outcomes following an injury. They further demonstrated that pre-existing mental health conditions accounted for almost half the mental health service use attributable to an injury. Likewise, a history of familial psychiatric problems is a consistent predictor of PTSD following injuries (Aupperle *et al.* 2012). Lin *et al.* (2014) claimed that life events before and after injury and previous occupational injury experiences are also risk factors. Personal health behaviours such as smoking, alcohol/substance dependence and lack of exercise are also cofounding covariates of SPIs among injured workers (Kim 2013).

Resilience factors

There are numerous factors that enhance resilience and protect workers against vulnerability to SPIs. These include pre-injury psychological traits, coping style, social support and workplace integration.

Dezarnaulds and Ilchef (2014) argued that the psychological response to injury sequelae is very much individual and mediated by pre-morbid characteristics, perceptions of the injury and self-management. Characteristics such as humour, optimism, a robust sense of self-efficacy and an ability to employ problem-solving to deal with difficulties help injured workers cope and adjust well. Coping refers to an individual's ongoing efforts in thought and actions to manage specific demands appraised as taxing his/her psychological well-being. Coping processes can be classified into two; problem-focused and emotion-focused. The problem-focused coping aims at problem solving or doing something to change the influence of the stressor. The emotion-focused coping aims at changing the way of attending to or interpreting what is happening, i.e. focusing on managing the emotional distress associated with the situation (Lazarus 1993). Using the two-factor model above, Carver *et al.* (1989) proposed 14 coping mechanisms, as follows:

1 Active coping – the process of taking active steps to remove or circumvent the stressor or to ameliorate its effects.
2 Planning – thinking and coming up with strategies to best handle the problem.
3 Suppression of competing activities – putting everything else aside in order to deal with the stressor.
4 Restraint coping – waiting until an appropriate opportunity to present itself, holding oneself back and not acting prematurely.
5 Social support for instrumental reasons – seeking advice, assistance or information.
6 Social support for emotional reasons – seeking moral support, sympathy or understanding.
7 Venting of emotions – focusing on the distress or upset an individual is experiencing, e.g. a period of mourning.
8 Behavioural disengagement – reducing the effort to deal with the stressor, even giving up the attempt to attain goals with which the stressor is interfering, particularly when poor coping outcomes are perceived.
9 Mental disengagement – using alternative activities or distractions to take the mind off a problem.
10 Positive reinterpretation or reappraisal and growth – construing a stressful situation in positive terms by, for example, looking for something good in what is happening.
11 Denial – refusing to believe that the stressor exists or trying to act as though the stressor is not real.
12 Acceptance – accepting the reality of a stressful situation.
13 Turning to religion – religion can serve as a source of emotional support, a vehicle for positive reinterpretation and growth, or a tactic of active coping.
14 Substance abuse – alcohol or drug dependence to distract the mind from the problem.

Studies agree that problem-focused coping is more effective than emotion-focused coping in warding off psychological injuries. Emotion-focused coping, however, can be effective when the stressor cannot be altered and during the immediate aftermath of the stressor (Carr and Umberson 2013). A meta-analysis of 36 relevant studies revealed that problem-focused coping correlated negatively with burnout symptoms whilst emotion-focused coping correlated positively. The study further elaborated that seeking social support, reappraisal, and religious copings were negatively correlated with burnout symptoms whilst acceptance was positively correlated (Shin *et al.* 2014).

Care, emotional support and acceptance from both family and peers act as buffers against the stress associated with trauma and facilitate resilience and adjustment post-injury (Sareen *et al.* 2013). Evidence suggests that injured workers, who are divorced or separated, are more vulnerable to SPIs (Cheadle *et al.* 1994). A soundly functioning marital relationship provides companionship, intimacy and social support, and also expands social networks, thus expanding the number of people who can be drawn on for assistance. These positive features of family relationship work to downplay the adverse effects of injury sequelae that workers face. Those unmarried or single individuals miss these comforters. In the case of separated, divorced or widowed workers, they have to deal with the trauma and life stress of relationship losses in addition to the work-injury induced stress, making them a more vulnerable group to SPIs.

Support from the employer to obtain full compensation and to integrate back into workforce appropriately moderates the psychological distress of work-injury induced disability or impairment. Allowing an adequate recovery period is also crucial because being forced back to work prematurely and having to work with pain elevate the risk of SPIs (Dembe 2001).

Construction and SPIs

The construction industry records one of the worst safety records among all industrial sectors globally. It is discernible that a large portion of construction workers who suffer from work-related injuries and illnesses are likely to suffer SPI consequences. Although construction is accident prone and its workers are exposed to SPI risk at a higher rate, studies of SPIs in construction are very limited. There is only anecdotal evidence with WorkCover Australia to suggest construction workers have claimed compensations for post-injury psychological disorders. SPIs is a largely under-explored area in construction. It is therefore imperative to develop mechanisms to prevent the creeping psychiatric consequences of work injuries and illnesses. Screening programmes administered soon after the injury could help identify the need for psychological treatments for injured workers alongside injury rehabilitation. Kendrick *et al.* (2011) specifically mentioned that screening tools may be useful in healthcare settings for identifying those at risk and early interventions are effective in treating psychological morbidity following injury. Lin *et al.* (2014) suggested that occupational injured workers at risk of SPIs can be identified by the examination of key injury and worker characteristics and be subjected to targeted screening and early intervention efforts. Accordingly, it can be postulated that:

- The likelihood of developing a SPI and potential severity can be predicted by the injury/illness characteristics such as the mechanism of the accident, type of injury, severity of physical injury/illness and body part affected, as well as construction worker characteristics such as occupation and demographics.

- Based on the predicted potential severity of SPI, construction workers at high risk may be identified and subjected to early intervention.

The next section operationalises these propositions using data analytics methods.

Neural network model for SPI severity prediction in construction

The previous section concluded that a predictive/screening tool may be developed to proactively identify occupationally injured construction workers who are at high risk of SPIs. This section demonstrates the development of the predictive tool, deploying the neural network (NN) approach to predictive analytics.

The human brain is an extremely complex, parallel information processing system, capable of learning from experiences/examples and then generalising the acquired knowledge for application in new cases, even with incomplete data (Ahiaga-Dagbui *et al.* 2013). Neural networks are a computational technique that imitate this quality of the human brain for information processing and problem solving. They are built using massively interconnected computational 'neurons' that perform the parallel-distributed information processing. The NN method incorporates many advanced qualities that are beneficial for information processing in the context of SPI severity prediction, which are:

- It is not necessary to know the solid underlying relationships between input and output variables for model building; any relationship whether linear or nonlinear can be learnt and approximated by the machine-learning quality of the technique (Thipparat 2012). This is particularly important for this research because to date no explanatory research has been conducted to establish causalities in SPI severities.
- Neural networks are able to learn complex nonlinear relationships between inputs and outputs from a dataset, obtained from field observations, and then serve as models of multivariate decision support systems (Babuška and Verbruggen 2003). They are suitable for modelling complex, hard-to-explain problems where no formal underlying theories or mathematical procedures are available (Adeli 2001).

The neural network technique involves complex mathematical algorithms and computations, and a discussion of them is beyond the scope of this book. Readers are referred to Karray and De Silva (2004). Figure 5.2 illustrates the process of developing a neural network model, which essentially involves six critical sequential steps: (1) identifying variables for modelling; (2) obtaining past data for model development; (3) preparing the data to make it suitable for modelling; (4) defining an initial neural network structure; (5) training the initial neural network using the training data; and (6) testing the performance of the final neural network model using the test dataset. The forthcoming sections shadow this systematic process flow to create the SPI severity prediction tool.

Figure 5.2 Neural network model development process.

Sources: adapted from Al Shamisi *et al.* 2011; Heravi and Eslamdoost 2015.

Identifying variables for SPI severity prediction

The literature review above discussed the risk factors associated with SPIs. Based on the propositions articulated above, a conceptual framework of SPI prediction is drawn, as illustrated in Figure 5.3. This guided the data collection and the subsequent steps of neural network model building.

Data collection and pre-processing

Data, containing 422 cases of SPIs that were recorded between July 2005 and August 2015, were obtained from the State Insurance Regulatory Authority of WorkCover New South Wales (NSW), Australia. A typical case was characterised by variables such as: date of primary physical injury, mechanism of incident, body part injured, nature/description of primary injury, time between the primary physical injury and SPI, gender of victim, age category of victim, occupation of victim, medical cost paid and weekly wages compensated. However, details of the severity of the primary physical injury suffered were not available in the dataset.

Pre-processing of the data was first undertaken to clean and prepare it for prediction modelling. Table 5.1 illustrates the pre-processing performed on the data. Thirty-two mechanisms of accidents in the dataset were grouped into seven (falls; hit by or trapped between objects; muscular stress; contact with electricity/heat/ substances; psychological stressor; vehicular accident; and other/unspecified mechanism). Fifty-five natures of primary injury were formed into ten clusters (fractures/

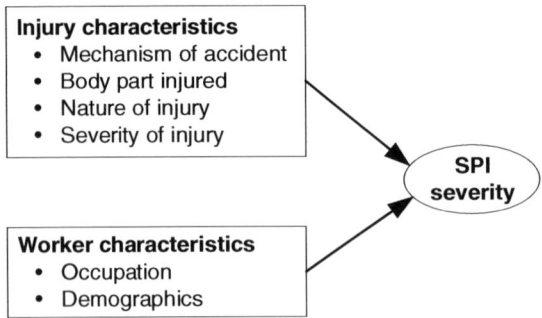

Figure 5.3 SPI severity prediction framework.

dislocations; muscle/joint/tendon/tissue damages; brain/nerves injuries; amputations; cuts, wounds and bruises; burns/electric shocks; spinal/back injuries; psychological disorders; multiple injuries; and other/unspecified injuries). Similarly, 51 body parts injured were clustered into nine categories (head and neck; shoulder; hand and upper limb; chest and abdomen; back; leg; psychological system; multiple; and other). Eleven age categories were reduced to six (below 20; 20–29; 30–39; 40–49; 50–59; and 60 and above). Occupation had three types: professional, tradesperson and labourer. The duration between the physical and psychological injuries had been recorded in days. However, this variable was not considered for modelling as it is more like an intermediate variable and the effect of it can be captured by the dependant variable that is discussed in the next paragraph. Input and output data for neural network modelling must be presented in numerical formats and numerical codes were therefore assigned to the categories above, as shown in column 3 of Table 5.1.

The dependent/response variable for the prediction is SPI severity. Both the medical cost and compensation amount paid to the victim reflect the severity of the SPI suffered. However, it is better to combine these two and have a single indicator of severity. A risk assessment matrix was utilised to combine and classify cases into one of the five risk ratings. Figure 5.4a illustrates individual ratings in a 1–5 scale against different cost groups and Figure 5.4b shows the combined ratings; the arithmetic sum of individual ratings provides the combined ratings (SPI severity scores). For each SPI case in the dataset, the medical cost and the wage compensation paid were first rated according to the scales shown in Figure 5.4a and then a combined score was computed, which was then used as the response variable.

Medical cost ($ range)	Severity rating
0 – 50,000	1
50,000 – 100,000	2
100,000 – 150,000	3
150,000 – 200,000	4
Over 200,000	5

Wage compensation ($ range)	Severity rating
0 – 50,000	1
50,000 – 100,000	2
100,000 – 150,000	3
150,000 – 200,000	4
Over 200,000	5

(a) Individual severity ratings

		Wage compensation severity rating				
		1	**2**	**3**	**4**	**5**
Medical cost severity rating	**5**	6	7	8	9	10
	4	5	6	7	8	9
	3	4	5	6	7	8
	2	3	4	5	6	7
	1	2	3	4	5	6

(b) Combined severity ratings

Figure 5.4 Severity rating.

Table 5.1 Input variables for prediction modelling

Variable name	Variable values	Numerical code	Count	Percentage
Mechanism of incident	Falls	1	129	30.6
	Hit by/trapped between objects	2	57	13.5
	Muscular stress	3	147	34.8
	Contact with electricity/heat/substances	4	11	2.6
	Vehicular accidents	5	24	5.7
	Psychological stressors	6	33	7.8
	Other/unspecified mechanisms	7	21	5.0
Nature of primary injury	Fractures/dislocations	1	62	14.7
	Muscle/joint/tendon/tissue damages	2	210	49.8
	Brain/nerves injuries	3	11	2.6
	Amputations	4	7	1.7
	Cuts, wounds and bruises	5	38	9.0
	Burns/electric shocks	6	12	2.8
	Spinal/back injuries	7	19	4.5
	Psychological disorders	8	34	8.1
	Multiple injuries	9	13	3.1
	Other/unspecified injuries	10	16	3.8

Body part injured			
	Head and neck	1	41	9.7
	Shoulder	2	35	8.3
	Hand and upper limb	3	36	8.5
	Chest and abdomen	4	12	2.8
	Back	5	127	30.1
	Leg	6	53	12.6
	Psychological system	7	34	8.1
	Multiple	8	82	19.4
	Other	9	2	0.5
Age of victim			
	Below 20	1	25	5.9
	20–29	2	78	18.5
	30–39	3	96	22.7
	40–49	4	116	27.5
	50–59	5	83	19.7
	60 and above	6	24	5.7
Occupation of victim			
	Professional	1	13	3.1
	Tradesperson	2	196	46.4
	Labourer	3	213	50.5

It is common in prediction modelling that the dataset is divided into at least two whereby one dataset is used for developing the model (also called as training) and the other dataset is used for testing the performance of the model (also known as validation). Accordingly, the dataset was randomly split into two: training set (392 cases) and testing set (30 cases). The size of the testing set was set to 30 to provide more cases for model training whilst ensuring a statistically valid sample size of the testing set.

Neural network model building

IBM SPSS Statistics 24 was deployed for building the neural network model for predicting SPI severities in the construction industry with the Multilayer Perceptron Model option chosen. After assigning variables for input and output layers of the model, the network architecture was defined with the following parameters:

- Number of hidden layers – 2
- Number of units/neurons in the hidden layer – automatically compute
- Activation function for the hidden layer – hyperbolic tangent
- Activation function for the output layer – hyperbolic tangent
- Type of training – online

With these parameters set, the supervised training of the neural network for predicting SPI severities was conducted. The outputs of model building by machine learning are provided below. Figure 5.5 illustrates the developed neural network model for SPI severity prediction in a simplified form and Table 5.2 shows its

Table 5.2 Neural network details

Input layer	Factors	1	Accident mechanism
		2	Injury
		3	Body part damaged
		4	Age
		5	Occupation
	Number of units[a]		35
Hidden layer(s)	Number of hidden layers		2
	Number of units in hidden layer 1[a]		10
	Number of units in hidden layer 2[a]		8
	Activation function		Hyperbolic tangent
Output layer	Dependent variables	1	SPI score
	Number of units		1
	Rescaling method for scale dependents		Adjusted normalised
	Activation function		Hyperbolic tangent
	Error function		Sum of squares

Note
a Excluding the bias unit.

Input layer   **Hidden layers**   **Output**

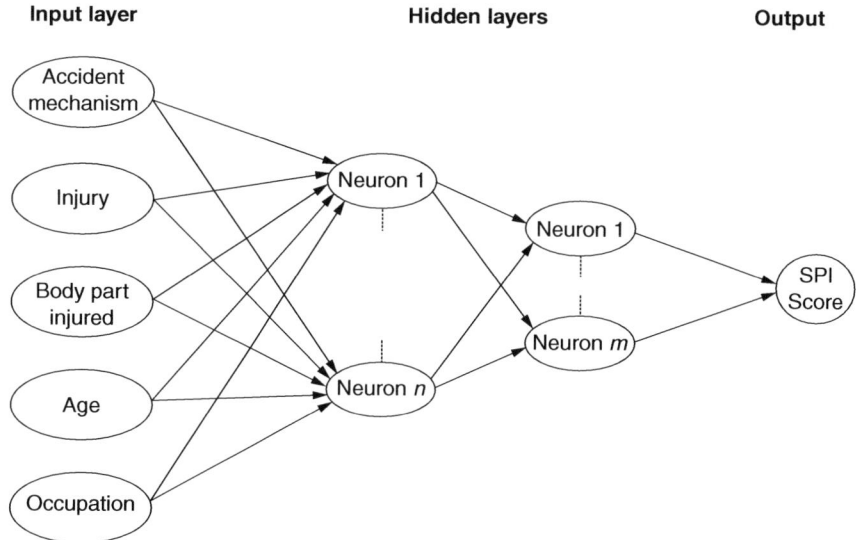

Figure 5.5 Neural network model structure.

details. The model has five input factors, automatically forming 35 units/neurons representing input categories. Similarly, hidden layers 1 and 2 consist of 10 and eight neurons, respectively.

The model summary (Table 5.3) displays information about the results of training the neural network. Sum-of-squares error is displayed because the output layer has a scale-dependent variable. This is the error function that the network tries to minimise during training. The relative error for each scale-dependent variable is the ratio of the sum-of-squares error for the dependent variable to the sum-of-squares error for the 'null' model, in which the mean value of the dependent variable is used as the predicted value for each case.

The predicted-by-observed chart (Figure 5.6) displays a scatterplot of predicted values of SPI Score on the *y* axis by observed values on the *x* axis for

Table 5.3 Model summary

Training	Sum of squares error	33.082
	Relative error	0.757
	Stopping rule used	1 consecutive step(s) with no decrease in error[a]
	Training time	0:00:00.13

Dependent Variable: SPI score

Note
a Error computations are based on the training sample.

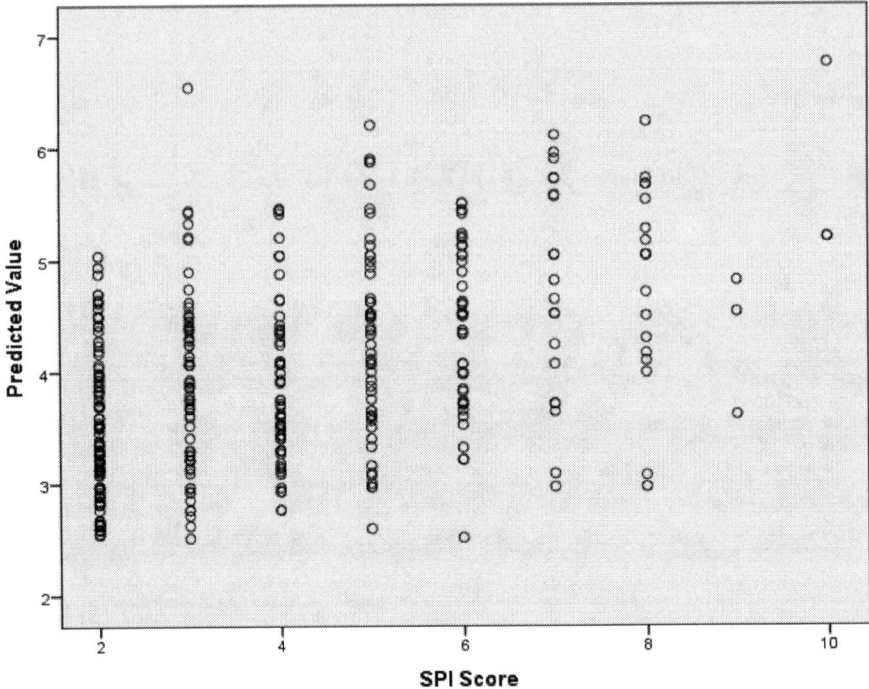

Figure 5.6 Predicted-by-observed chart.

the training samples. Ideally, values should lie roughly along a 45-degree line starting at the origin. The points in this plot form vertical lines at each observed value SPI score. Looking at the plot, it appears that the model does a reasonably good job of predicting SPI scores. The general trend of the plot is off the ideal 45-degree line, which suggests that predictions for observed SPI scores of under four tend to overestimate, while predictions for observed SPI scored beyond six tend to underestimate.

The residual-by-predicted chart (Figure 5.7) displays a scatterplot of the residual (observed value minus predicted value) on the y axis by the predicted value on the x axis. Each diagonal line in this plot corresponds to a vertical line in the predicted-by-observed chart, and the progression from over-prediction to under-prediction can be seen clearly as the observed SPI score increases.

The importance chart (Figure 5.8) shows that the prediction results are dominated by the body part damaged, followed by the nature of injury suffered, followed closely by the occupation of the victim. The weight of the mechanism of accident and the age of the victim in the SPI score prediction is significantly lower than the above three variables.

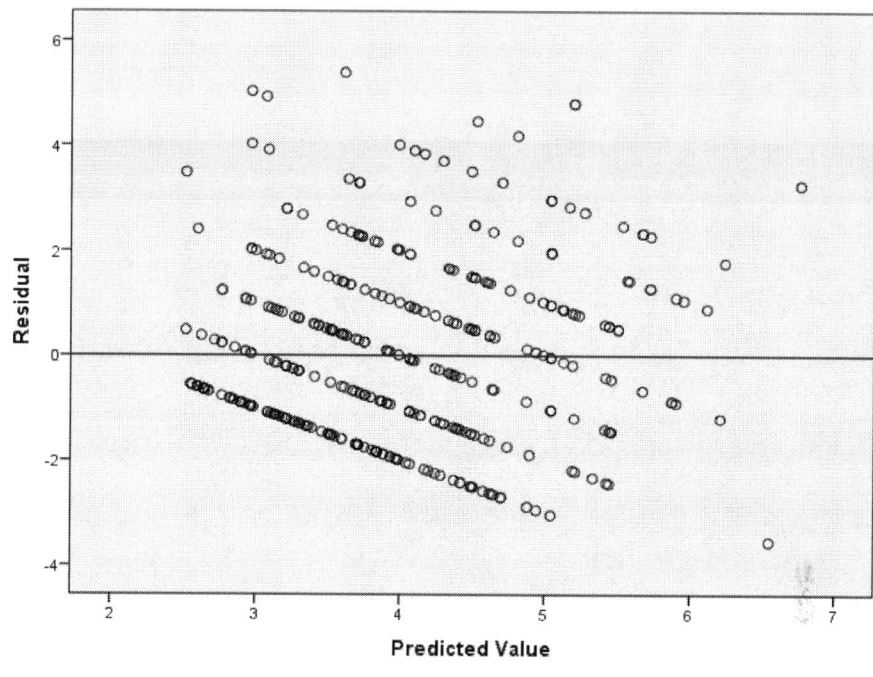

Dependent Variable: SPI Score

Figure 5.7 Residual-by-predicted chart.

Model performance

The model's performance for predicting SPI severities was evaluated using the test dataset, which had 30 known cases. The values for accident and employee characteristics that had been recorded in the cases were used to predict the SPI severity scores using the neural network model and then the results were compared with the actual SPI severity scores recorded in the test dataset. Comparison results are shown in Table 5.4. Columns 1 to 5 represent numerical codes for input values as designated in Table 5.1. Column 6 contains the actual SPI score whilst column 7 depicts model predicted values. Column 8 displays a prediction error category for each case, which were derived based on the following classifications:

- Score different ≤1.00 = Very low prediction error
- 1.00 < score difference ≤2.00 = Low prediction error
- 2.00 < score difference ≤3.00 = Moderate prediction error
- 3.00 < score difference ≤4.00 = High prediction error
- Score different >4.00 = Very high prediction error

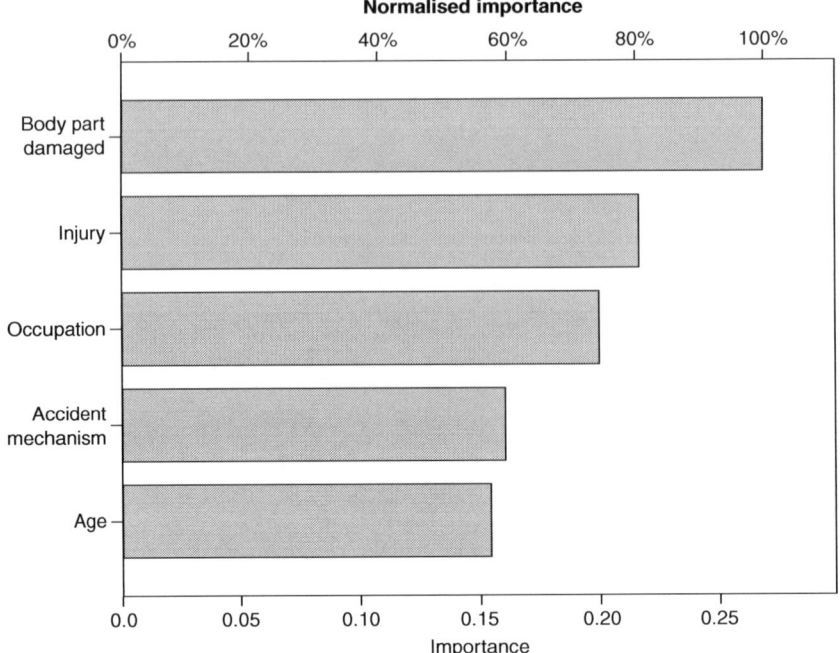

Figure 5.8 Independent variable importance chart.

According to these error classifications, the predicted values yielded the following performance outcomes:

- Very low prediction error = 14 cases
- Low prediction error = 9 cases
- Moderate prediction error = 2 cases
- High prediction error = 5 cases
- Very high prediction error = none

It can be concluded based on the results that 23 out of 30 (77 per cent) had very low or low prediction errors and only five out of 30 cases (17 per cent) had serious prediction errors. Hence, the developed model can be deemed as a good predictor of SPI severities among construction injury victims.

The neural network model has three key advanced qualities that are not present in other prediction modelling methods. First, the prediction algorithm in the model was derived by machine learning from real-world data collected by the State Insurance Regulatory Authority of WorkCover New South Wales (NSW) and as such reflects the true situation of the NSW construction industry. Second, the system has a good prediction accuracy. Finally, it is easy to update the model

Table 5.4 Neural network model performance

C1	C2	C3	C4	C5	C6	C7	C8
Accident mechanism	Injury	Body part damaged	Age	Occupation	Actual SPI Score	ANN predicted value	Prediction Error category
3	2	5	3	3	2	3.71	Low
2	1	4	4	3	4	3.44	Very low
3	2	5	3	2	5	4.03	Very low
5	2	8	6	2	6	4.79	Low
1	3	1	5	2	3	6.59	High
1	3	1	4	3	7	3.76	High
3	2	2	4	2	8	4.19	High
6	8	7	2	1	2	2.14	Very low
3	2	5	3	2	3	4.03	Very low
1	2	8	2	2	4	4.52	Very low
3	2	8	3	2	5	4.12	Very low
1	2	5	5	3	4	3.52	Very low
3	2	5	4	2	3	4.55	Low
3	7	5	3	2	7	5.56	Low
2	5	3	3	3	2	3.93	Low
2	1	2	5	2	6	4.68	Low
1	1	2	5	3	3	3.87	Very low
4	6	3	4	3	2	3.00	Low
1	1	3	3	3	5	3.86	Low
3	2	5	6	2	4	4.05	Very low
5	2	1	2	2	5	4.22	Very low
7	6	8	4	2	2	5.80	High
3	7	5	4	3	6	3.33	Moderate
5	2	5	2	3	3	3.25	Very low
1	1	4	4	3	3	3.51	Very low
1	2	8	3	3	2	4.08	Moderate
1	9	8	3	3	3	3.88	Very low
1	2	1	3	3	5	4.04	Very low
7	2	8	2	2	9	5.47	High
7	2	5	5	3	4	2.95	Low

for other contexts and future changes since machine learning drives the development of the prediction algorithm. The neural network model is easily adaptable in the future with minimal effort as new cases can be input for re-learning and automatically updating the algorithm, providing the neural network method of prediction with longer application and improved accuracy capabilities.

Limitations

The literature review identified four accident-related variables, such as mechanism of accident, type of injury, severity of injury and body part injured, and two employee-related variables such as occupation and demographics (age and gender). However, the prediction model presented above did not include two variables, namely: severity of injury and gender for the following reasons:

- Details about the severity of physical injury suffered by the employees were not available in the dataset provided by the State Insurance Regulatory Authority of WorkCover New South Wales (NSW), Australia.
- Gender was not used as one of the predictor variables as there was not a representative number of female employees in the dataset; only five females out of 422 employees. Thus, incorporating gender in the modelling would not have changed the results, although the data related to females were included in model building.

In the same vein, variables related to resilience and personal psychiatric history of individuals were not included in the modelling as there were no records in the database for the variables. Another limitation is the size of the dataset. The prediction modelling involved a training dataset of 392 cases and it would have been better to use a larger dataset for improved accuracy. However, this is dictated by the availability of real-world data. Despite these limitations, the neural network model developed proved to be reasonably accurate in its prediction of SPI severities.

Conclusion

This study applied data analytics methodologies on workers' compensation claims data for SPIs to develop a predictive screening tool that enables healthcare providers and rehabilitation programme facilitators to proactively identify potential SPI sufferers and thereby initiate early intervention measures. This will curtail numerous long-term socio-economic consequences and human suffering associated with work-related psychological disorders.

This research contributes to the practices in public health, work health and safety and workers' compensation insurance fields. Most countries have legislations, policies and systems that regulate workers' compensation insurance payments, rehabilitation and return-to-work programmes and the provision of medical services for occupational injured workers. Nonetheless, their focus is predominantly on the primary injury or illness and its subsequent impacts on the mental health do not receive adequate attention despite it having been established that the aftereffects of work injuries and illnesses have severe negative impacts on the workers, their families and even employers. To this end, the following are suggested to proactively prevent SPIs among construction workers:

- Treatment and healthcare services for occupational injured workers should incorporate diagnosis for mental health issues as a compulsory part in the initial and follow up treatments. The diagnosis should correspond to the length of the impairment or disability suffered by the worker. Many injured workers do not voluntarily report or seek mental health services for various reasons, resulting in the absence of medical evidence. The absence of medical confirmation of mental health limits access to care, welfare support and workers' compensation benefits, potentially sustaining the vicious cycle of SPI and injury/impairment/disability. A compulsory diagnosis will enable workers to access the socio-economic and medical support duly.

- Rehabilitation and return-to-work programmes should incorporate psychological therapies within their programmes and provide equal importance to them just like physiotherapy and similar. Psychological interventions/ treatments such as cognitive behaviour therapy (CBT), interpersonal psychotherapy (IPT), mindfulness-based cognitive therapy, positive psychology, psychotherapies, counselling and narrative therapy may be integrated into the rehabilitation programme.
- Work health and safety legislations and policies largely focus on preventing physical injuries and illnesses at workplaces. Provisions within the legislations are essential that designate responsibilities and accountability of employers for risk management of SPIs just like for physical hazards on work sites.
- Workers' compensation insurance providers traditionally decide their premium amount for insurance policies by primarily analysing past claims data filed by work injury victims. Given the fact that about 50 per cent of injury victims suffer subsequent SPIs, the actual compensation to be paid will be significantly more than the physical injury claim and therefore the actual risk assumed by the insurer. Insurance providers may factor SPIs into their premium computation algorithms to reduce their risks.

This research contributes to the construction management body of knowledge in three different ways. First, the study has introduced a new dimension of workplace injuries, secondary psychological injuries, into the construction health and safety literature and thus sows the seeds for future research by others in this critical, growing field. Second, the study has produced the prototype of a neural network model for predicting SPI severities in construction. The model is readily usable by construction organisations, healthcare providers and rehabilitation programme facilitators to proactively identify potential SPI sufferers and thereby initiate early intervention measures. Finally, there are methodological insights for construction safety researchers. Despite the increased use of neural networks in many fields, to date it is largely an overlooked method in construction health and safety research, and specifically in studies of psychological injuries and work stress, possibly due to the lack of previous research that showcases the application effectiveness in this area of construction research. This research therefore provides fresh insights for construction health and safety researchers and serves as an exemplary application.

References

Adeli, H. (2001). Neural networks in civil engineering: 1989–2000. *Computer-Aided Civil and Infrastructure Engineering*, 16(2): 126–142.

Ahiaga-Dagbui, D.C., Tokede, O., Smith, S.D., and Wamuziri, S. (2013). A neuro-fuzzy hybrid model for predicting final cost of water infrastructure projects. In S.D. Smith and D.D. Ahiaga-Dagbui (Eds), Proceedings of the 29th Annual ARCOM Conference, 2–4 September 2013, Reading, UK. ARCOM, 181–190.

Aitken, L.M., Chaboyer, W., Kendall, E., and Burmeister, E. (2012). Health status after traumatic injury. *Journal of Trauma and Acute Care Surgery*, 72(6): 1702–1708.

Al Shamisi, M.H., Assi, A.H., and Hejase, H.A.N. (2011). Using MATLAB to develop artificial neural network models for predicting global solar radiation in Al Ain City – UAE. Available online: http://cdn.intechopen.com/pdfs/21382.pdf (accessed 8 May 2017).

American Psychiatry Association (APA) (1994). *Diagnostic and Statistical Manual of Mental Disorder*, 4th edn (DSM-IV) (pp. 427–429). Washington, DC: American Psychiatry Association.

Asfaw, A. and Souza, K. (2012). Incidence and cost of depression after occupational injury. *Journal of Occupational & Environmental Medicine*, 54(9): 1086–1091.

Aupperle, R.L., Melrose, A.J., Stein, M.B., and Paulus, M.P. (2012). Executive function and PTSD: Disengaging from trauma. *Neuropharmacology*, 62(2): 686–694.

Babuška, R. and Verbruggen, H. (2003). Neuro-fuzzy methods for nonlinear system identification. *Annual Reviews in Control*, 27: 73–85

Bay, E. and Donders, J. (2008). Risk factors for depressive symptoms after mild-to-moderate traumatic brain injury. *Brain Injury*, 22(3): 233–241.

Beyondblue (n.d.). Serious injury and anxiety, depression and post-traumatic stress disorder. Available online: http://resources.beyondblue.org.au/prism/file?token=BL/0838 (accessed 12 February 2017).

Braverman, M. (1992). Post-trauma crisis intervention in the workplace. In J.C. Quick, L.R. Murphy and J.J. Hurrell (Eds), *Stress and Well-being at Work: Assessment and Interventions for Occupational Mental Health* (pp. 299–316). Washington, DC: APA Press.

Cameron, C.M., Purdie, D.M., Kliewer, E.V., and McClure, R.J. (2006). Mental health: A cause or consequence of injury? A population-based matched cohort study. *BMC Public Health*, 6: 114.

Carr, D. and Umberson, D. (2013). The social psychology of stress, health, and coping. In J. DeLamater and A. Ward (Eds), *Handbook of Social Psychology*, 2nd edn. Dordrecht, the Netherlands: Springer.

Carver, C.S., Scheier, M.F., and Weintraub, J.K. (1989). Assessing coping strategies: A theoretically based approach. *Journal of Personality and Social Psychology*, 56(2): 267–283.

Chang, C.M., Connor, K.M., Lai, T.J., Lee, L.C., and Davidson, J.R.T. (2005). Predictors of post-traumatic outcomes following the 1999 Taiwan earthquake. *The Journal of Nervous and Mental Disease*, 193(1): 40–46.

Cheadle, A., Franklin, G., Wolfhagen, C., Savarino, J., Liu, P.Y., Salley, C., and Weaver, M. (1994). Factors influencing the duration of work-related disability: A population-based study of Washington state workers' compensation. *American Journal of Industrial Medicine*, 34(2): 190–196.

Cheung, E., Alvaro, R., and Colotla, V.A. (2003). Psychological distress in workers with traumatic upper or lower limb amputations following industrial injuries. *Rehabilitation Psychology*, 48(2): 109–112.

Dawson, S.E. (1994). Workers' compensation in Pennsylvania: The effects of delayed contested cases. *Journal of Health and Social Policy*, 6(1): 87–100.

Dembe, A.E. (1998). Evaluating the impact of managed health care in workers' compensation. In: J. Harris (Ed.), *Managed Care* (pp. 134–156). Philadelphia, PA: Hanley & Belfus.

Dembe, A.E. (2000). Pain, function, impairment and disability: Implications for workers' compensation and other disability insurance systems. In: T. Mayer, R. Gatchel, and P. Polatin (Eds), *Occupational Musculoskeletal Disorders: Function, Outcomes, and Evidence* (pp. 563–576). New York: Lippincott, Williams & Wilkins.

Dembe, A.E. (2001). The social consequences of occupational injuries and illnesses. *American Journal of Industrial Medicine*, 40(4): 403–417.

Dezarnaulds, A and Ilchef, R. 2014. *Psychological Adjustment after Spinal Cord Injury.* Chatswood, NSW: Agency for Clinical Innovation (ACI). www.aci.health.nsw.gov.au/__data/assets/pdf_file/0010/155197/Psychosocial-Adjustment.pdf (24 April 2019).

Esselman, P.C., Askay, S.W., Carrougher, G.J., Lezotte, D.C., Holavanahalli, R.K., Magyar-Russell, G., Fauerbach, J.A. and Engrav, L.H. (2007). Barriers to return to work after burn injuries. *Archives of Physical Medicine and Rehabilitation*, 88(12 Suppl 2): S50–S56.

Heravi, G. and Eslamdoost, E. (2015). Applying artificial neural networks for measuring and predicting construction labor productivity. *Journal of Construction Engineering and Management*, 141(10): 0001006–04015032.

Karray, F.O. and De Silva, C. (2004). *Soft Computing and Intelligent Systems Design: Theory, Tools and Applications.* Essex: Pearson Education Limited.

Kellezi, B., Coupland, C., Morriss, R., Beckett, K., Joseph, S., Barnes, J., Christie, N., Sleney, J., and Kendrick, D. (2016). The impact of psychological factors on recovery from injury: a multicentre cohort study. *Social Psychiatry and Psychiatric Epidemiology*, doi:10.1007/s00127-016-1299-z.

Kendrick, D., O'Brien, C., Christie, N., Coupland, C., Quinn, C., Avis, M., Barker, M., Barnes, J., Coffey, F., Joseph, S., Morris, A., Morriss, R., Rowley, E., Sleney, J., and Towner, E. (2011). The impact of injuries study. Multicentre study assessing physical, psychological, social and occupational functioning post injury – a protocol. *BMC Public Health*, 11: 963.

Keogh, J., Nuwayhid, I., Gordon, J., and Gucer, P. (2000). The impact of occupational injury on injured worker and family: Outcomes of upper extremity cumulative trauma disorders in Maryland workers. *American Journal of Industrial Medicine*, 38(5): 498–506.

Kim, J. (2013). Depression as a psychosocial consequence of occupational injury in the US working population: Findings from the medical expenditure panel survey. *BMC Public Health*, 13: 303.

Krause, N., Frank, J., Dasinger, L., Sullivan, T.J., and Sinclair, S.J. (2001). Determinants of duration of disability and return-to-work after work-related injury and illness: Challenges for future research. *American Journal of Industrial Medicine*, 40(4): 464–484.

Kuo, H.W., Wu, S.J., Ma, T.C., Chiu, M.C., and Chou, S.Y. (2007). Post-traumatic symptoms were worst among quake victims with injuries following the Chi-Chi quake in Taiwan. *Journal of Psychosomatic Research*, 62: 495–500.

Lazarus, R.S. (1993). From psychological stress to the emotions: A history of changing outlooks. *Annual Review of Psychology*, 44: 1–22.

Lin, K.H.1, Shiao, J.S., Guo, N.W., Liao, S.C., Kuo, C.Y., Hu, P.Y., Hsu, J.H., Hwang, Y.H., and Guo, Y.L. (2014). Long-term psychological outcome of workers after occupational injury: Prevalence and risk factors. *Journal of Occupational Rehabilitation*, 24(1): 1–10.

MacDonald, H.A., Colotla, V., Flamer, S., and Karlinsky, H. (2003). Posttraumatic stress disorder (PTSD) in the workplace: A descriptive study of workers experiencing PTSD resulting from work injury. *Journal of Occupational Rehabilitation*, 13(2): 63–77.

Mason, S., Turpin, G., Woods, D., Wardrope, J., and Rowlands, A. (2006). Risk factors for psychological distress following injury. *British Journal of Clinical Psychology*, 45(Pt 2): 217–230.

Morse, T., Dillon, C., Warren, N., Levenstein, C., and Warren, A. (1998). The economic and social consequences of work-related musculoskeletal disorders: The Connecticut Upper-extremity Surveillance Project (CUSP). *International Journal of Occupational and Environmental Health*, 4: 209–216.

O'Donnell, M.L., Creamer, M., Pattison, P., and Atkin, C. (2004). Psychiatric morbidity following injury. *American Journal of Psychiatry*, 161(3): 507–514.

O'Donnell, M.L., Creamer, M., Holmes, A.C., Ellen, S., McFarlane, A.C., Judson, R., Silove, D., and Bryant, R.A. (2010). Posttraumatic stress disorder after injury: Does admission to intensive care unit increase risk? *Journal of Trauma*, 69(3): 627–632.

O'Hagan, F.T., Ballantyne, P.T., and Vienneau, P. (2012). Mental health status of Ontario injured workers with permanent impairments. *Canadian Journal of Public Health*, 103(4): e303–e308.

Pransky, G., Benjamin, K., Hill-Fatouhi, C., Himmelstein, J., Fletcher, K., Katz, J., and Johnson, W. (2000). Outcomes in work-related upper extremity and low back injuries: Results of a retrospective study. *American Journal of Industrial Medicine*, 37(4): 400–409.

Reville, R.T. (1999). The impact of a disabling workplace injury on earnings and labor force participation. In: J. Haltiwanger, J. Lane, J.R. Spletzer, J. Theeuwes, and K. Troske (Eds), *The Creation and Analysis of Employer–Employee Matched Data*. Amsterdam: Elsevier Science.

Sareen, J., Erickson, J., Medved, M.I., Asmundson, G.J.G., Enns, M.W., Stein, M., Leslie, W., Doupe, M., and Logsetty, S. (2013). Risk factors for post-injury mental health problems. *Depression and Anxiety*, 30: 321–327.

Shih, R.A., Schell, T.L., Hambarsoomian, K., Belzberg, H., and Marshall, G.N. (2010). Prevalence of posttraumatic stress disorder and major depression after trauma center hospitalization. *Journal of Trauma*, 69: 1560–1566.

Shin, H., Park, Y.M., Ying, J.Y., Kim, B., Noh, H., and Lee, S.M. (2014). Relationships between coping strategies and burnout symptoms: A meta-analytic approach. *Professional Psychology: Research and Practice*, 45(1): 44–56.

Strunin, L. and Boden, L. (1997). The human costs of occupational injuries. Presentation at the National Occupational Injury Research Symposium, Morgantown, West Virginia, 15–17 October 1997.

Sullivan, V.C. (2004). *Understanding Mental Illness*. Kanahooka, NSW: Betta Binding and Printing.

Tarasuk, V. and Eakin, J.M. (1995). The problem of legitimacy in the experience of work-related back injury. *Qualitative Health Research*, 5(2): 204–221.

Thipparat, T. (2012). Application of adaptive neuro fuzzy inference system in supply chain management evaluation. http://cdn.intechopen.com/pdfs/34230.pdf (accessed 29 March 2016).

WorkCover Queensland (2015). *Spotting the Secondary Psychological Injury Red Flags*. Available from: www.worksafe.qld.gov.au/health/articles/spotting-the-secondary-psychological-injury-red-flags (accessed 30 November 15).

WorkCover South Australia (2012). Managing psychological injuries: A guide for rehabilitation and return to work coordinators. http://library.safework.sa.gov.au/attachments/58834/Managing%20psychological%20injuries%20for%20rehabilition%20and%20return%20to%20work%20coordinators.pdf (accessed 22 March 2016).

Yang, Y.K., Yeh, T.L., Chen, C.C., Lee, C.K., Lee, I.H., Lee, L.C. and Jeffries, K.J. (2003). Psychiatric morbidity and posttraumatic symptoms among earthquake victims in primary care clinics. *General Hospital Psychiatry*, 25: 253–261.

6   Conclusion

This chapter concludes the book by producing a new, comprehensive causation model for incidents in the construction industry, drawing from the findings discussed in the previous chapters. It further discusses the practical implications of the proposed new causation model, followed by new methodological insights demonstrated in this book. Finally, the chapter outlines some future research potentials in the domain of improving safety and well-being in construction through data mining and analytics methods.

Summary of research and findings

The construction industry is one of the most vulnerable sectors for workplace incidents globally; it employs only 7 per cent of the global workforce but accounts for 30–40 per cent of work fatalities. One worker is killed every five minutes on a construction site worldwide. The construction industry is also notorious for health issues and high levels of work stress and related psychological diseases. The unacceptably high rate of incidents in construction causes a distressing socio-economic burden for the victims, their families and friends, employers and society in general, globally. Improving safety and well-being in construction is therefore an urgent need to save lives and the socio-economic well-being of nations.

Health and safety authorities collect enormous amount of data on construction workplace incidents every year. Construction researchers advocate that these past incident records are laden with the potential for learning lessons and creating new knowledge for improving health, safety and well-being. Researchers in other fields have demonstrated that data mining and analytics methods on past workplace incident records lent themselves to discover new patterns and insights to improve safety, which were not possible with traditional analysis techniques. However, it is an underexplored area for construction. To this end, this book aimed to harness the power of data mining and analytics techniques to discover new insights to improve health, safety and well-being in construction. The techniques were applied on past incident data that were obtained from Safe Work Australia to expand the understanding and knowledge in four different themes, namely: curtailing fatalities, addressing challenges with work-induced diseases, understanding the causes for psychological disorders in construction, and preventing secondary psychological injuries in construction. The key findings of these studies are summarised below.

Fatalities in construction

Current literature on construction fatalities suggests a notion, called the 'fatal four', which identifies four incident mechanisms that lead to fatalities among construction workers. These are: falls, struck-by object incidents, electrocution, and caught in/between objects incidents. However, the study discussed in Chapter 2 discovered six patterns of fatal incidents, and the agents of fatalities discovered are: falls, powered tools, plant and machinery, mental stress, noise, and chemical and other substances. The study further discovered the associations among the characteristics related to workers and work activities within each cluster.

The existing literature on construction falls relates fatalities to fall heights and the agents involved (i.e. scaffolding, ladders, elevated platforms and roofs). This study reveals that only falls that cause serious damages to the neck or head as well as incidents that involved older workers result in fatalities. Powered tools such as nail guns, drills, various saws, jack hammers, wood chippers and sanders are predominantly involved in construction fatalities, caused by struck-by broken/spinning parts of the tool or electrocution. An electric shock can also trigger falls from elevated platforms, resulting in death. Plant and machinery such as cranes, trucks and excavators are linked with fatalities due to struck-by, run over or overturning incidents on construction sites. Poor mental health among construction workers and apprentices aged below 24 as well as professionals has been a trigger for suicide-related fatalities in construction. Noise acts as an intermediary for fatalities in construction. Exposure to continuous excessive noise causes subtle hearing loss in workers, making alarms and warning signals from moving plant and machinery difficult to hear. Moreover, noise can lead to cardiovascular diseases and related deaths. Operators of impact equipment and tools (e.g. piling hammers, concrete breakers, manual hammers) and users of explosives (e.g. blasting, cartridge tools) are more vulnerable to fatalities of this nature. Exposure to chemicals and other pollutants, particularly asbestos, on construction sites can cause cancer and respiratory system diseases, eventually resulting in deaths, predominantly among workers aged 60 and above. The occupation types that are vulnerable to this risk include: metal workers, painters, repair trades and roofers.

Work-induced diseases in construction

The incidents of work-induced diseases are as prevalent as work injuries in the construction industry but pose more challenges for operatives. That is, in addition to causing socio-economic sufferings, such as injuries and fatalities, obtaining fair workers' compensation for work-induced diseases is often confronted with uncertainties because of the difficulties in proving work-related exposure or causes for the disease. This challenge can be curtailed with the establishment of analytics-based evidence that shows association patterns among the work-induced diseases, worker and work characteristics in the construction industry. The most common work-induced diseases suffered by construction operatives are musculoskeletal disorders and noise-induced hearing loss. These two diseases account for 96 per cent of the permanent disabilities suffered due to occupational diseases, whilst cardiovascular diseases and cancer are responsible for almost two-thirds of fatalities due to occupational diseases. The study discussed in Chapter 3 discovered the association and causal patterns for these four diseases.

Construction work is physically demanding, with exposure to forceful exertions, awkward body positions, bending and twisting of the body, repetitive motions, whole body vibration and work in kneeling, stooping, squatting positions. As a result, musculoskeletal disorders (MSDs) among construction workers is rampant and the top ten sufferers are identified as follows: carpenters, plumbers, electricians, structural steel construction workers, concreters, building labourers, truck drivers, plasterers, painters, and earth moving plant operators. MSDs largely cause temporary incapacities but at times they can lead to permanent disabilities for certain occupational trades. Accordingly, carpenters and building labourers appear to have suffered permanent disabilities due to MSDs.

Operations on construction sites are noisy, and long-term exposure to the noise can cause permanent hearing losses regardless of the occupation type as the noise can affect trades who are involved in noisy activities as well as others who work in the surroundings. Workers of any age group subject to long-term noise exposure on construction sites are susceptible to hearing loss. It is the length of employment in construction rather than the age of workers that is relevant.

As for cardiovascular diseases, construction operatives such as building labourers, electricians, bricklayers/stonemasons and metal fitters/machinists are vulnerable to fatalities due to ischaemic heart disease (heart attack) and other coronary diseases that cause blockages in the arteries of the heart. Moreover, workers age 50 and above in the aforementioned trades are more susceptible to heart attacks.

When analysing cancer incidents, two types of cancer, namely skin cancer and mesothelioma (cancer of lungs, abdomen, heart and testicles) are predominant among construction operatives in Australia. Skin cancer is reported to be caused by exposure to direct sun while on site. Long-term exposure to chemicals or substances and non-ionising radiation are responsible for mesothelioma in Australia. Workers aged 50 and above in occupations such as plumber, electrician, concreter, insulation installer, wall and floor tiler, plasterer, structural steel construction worker, plant operator and labourer are more vulnerable to these cancers.

Certain diseases may not appear during the course of occupation or soon after the exposure, but may take several years, even decades. Hence, latency periods are to be considered for certain fatal diseases, as follows: ischemic heart diseases – 10 to 30 years; mesothelioma – 10 to 50 years; skin cancer – two years; and musculoskeletal disorders – up to 12 years.

Psychological injuries in construction

The construction industries in many countries are known for work stress and consequent psychological issues, including suicides, among both operatives and professionals. The nature of stressors faced, the level of stress borne, and the severity of the psychological ill-health suffered vary across different occupation groups. Chapter 4 of this book explored the key stressors encountered by different worker groups and their impact on the level of work stress and consequent psychological ill-health suffered.

Managers in the construction industry are faced with multiple stressors simultaneously, which make them suffer adjustment disorders. The co-occurring stressors include: time pressure, working long hours, volume of paper work, staff shortage, responsibility for situation not fully in one's control, insufficient time to pursue leisure interests, insufficient time spent with family/home, travel to and from the job and inadequacy of communication flow. Construction professionals who constantly encounter high work pressure, workplace harassment, bullying and/or poor physical work environment are likely to suffer psychological injuries requiring days off work from 15 to 365 days, regardless of the size of the organisation they are employed by. Technicians, trade workers, machinery operators, drivers and labourers suffer mental ill-health largely due to being exposed to traumatic events, which could be witnessing a fatal accident of a co-worker, sudden collisions or toppling of machinery they operate. This could lead to post-traumatic stress disorders. Suicide or attempted suicide is prevalent among young workers, aged 20 and below and with the lowest weekly income. They are more likely to be apprentices. However, the primary stressor that pushes them to take the extreme action is bullying in the workplace. The representation of females is minimal in the construction workforce, yet they are not immune to psychological issues. The primary stressors encountered by them on construction sites are sexual harassment and/or gender discrimination.

Secondary psychological injuries in construction

Medical evidence suggests that workers who suffer serious workplace injuries or illnesses have a high likelihood of developing subsequent psychological disorders due to the unfavourable changes occurring to their employment and socio-economic conditions. This is referred to as secondary psychological injuries (SPIs). Construction being one of the most incident-prone industries, the number of workers who are likely to suffer SPIs is high. Chapter 5 expounded on the development of a neural network based predictive tool to proactively identify injured construction workers who are at high risk of developing SPIs.

The likelihood of the onset and the severity of SPIs to be suffered by an injured worker is predictable with information related to the body part damaged, nature of injury, occupation of the victim, incident mechanism and the age of worker. Injuries/illnesses that severely affect the physical appearance and/or the functionality of the victim permanently elevate the risk of SPIs. Victims of fall-related injuries are more susceptible to SPIs. Having been in an intensive care unit for treatment is also a predictor of SPIs. Age has a positive correlation with SPIs. Occupation type is not a predictor, but income level and security on the continuity of income are predictors.

Despite it having been well-established that work injuries and illnesses may result in subsequent mental disorders, the existing workers' compensation systems in many countries do not cover SPIs, thus leaving affected workers and their families helpless. Since the construction industry is highly incident-prone and the proportion of injured workers who suffer SPIs is high, governments and WHS authorities should consider revising the workers' compensation system to also include SPIs.

New incident causation model for construction

The studies discussed in the preceding chapters of the book enable the expansion of existing incident causation theories. Since construction is one of the most dangerous industries, safety has been an important topic for research, with the causes of construction incidents explored extensively by various researchers, resulting in several incident causation theories, such as the Domino Theory (Heinrich 1931), Multiple Causation Model (Petersen 1971), Human Error Model (Peterson 1982), the Swiss Cheese Model (Reason 1990), Accident Root Cause Tracing Model (Abdelhamid and Everett 2000), Modified Statistical Triangle of Accident Causation Model (Carter and Smith 2006), and Hierarchy of Causal Influences Model (Gibb *et al.* 2006). These theories attempt to link a string of events that eventually lead to a final outcome, physical injury or death. Another set of theories in construction posits the psychological conditions of workers as a cause of physical accidents (for example: Leung *et al.* 2010; Larsson *et al.* 2008; Siu *et al.* 2004). The findings of this research extend these existing theories by incorporating two additional components: (1) work-induced psychological injuries; and (2) mental health impact of work injuries and illnesses, referred to as secondary psychological injuries. The next paragraph demonstrates this in the context of a widely-adopted incident causation model.

The Hierarchy of Causal Influences Model of Gibb *et al.* (2006) (see Figure 6.1) has been adopted by several researchers and WHS authorities to investigate construction incidents, suggesting that it is a comprehensive theory to explain the causes of construction incidents. The model postulates that incidents occur due to the shortfalls of the work team (actions, behaviours, capabilities, communications), workplace (layout/space, lighting/noise, hot/cold/wet, local hazards), materials (suitability, usability, condition) and equipment (suitability, usability, condition), which are referred to as 'immediate accident

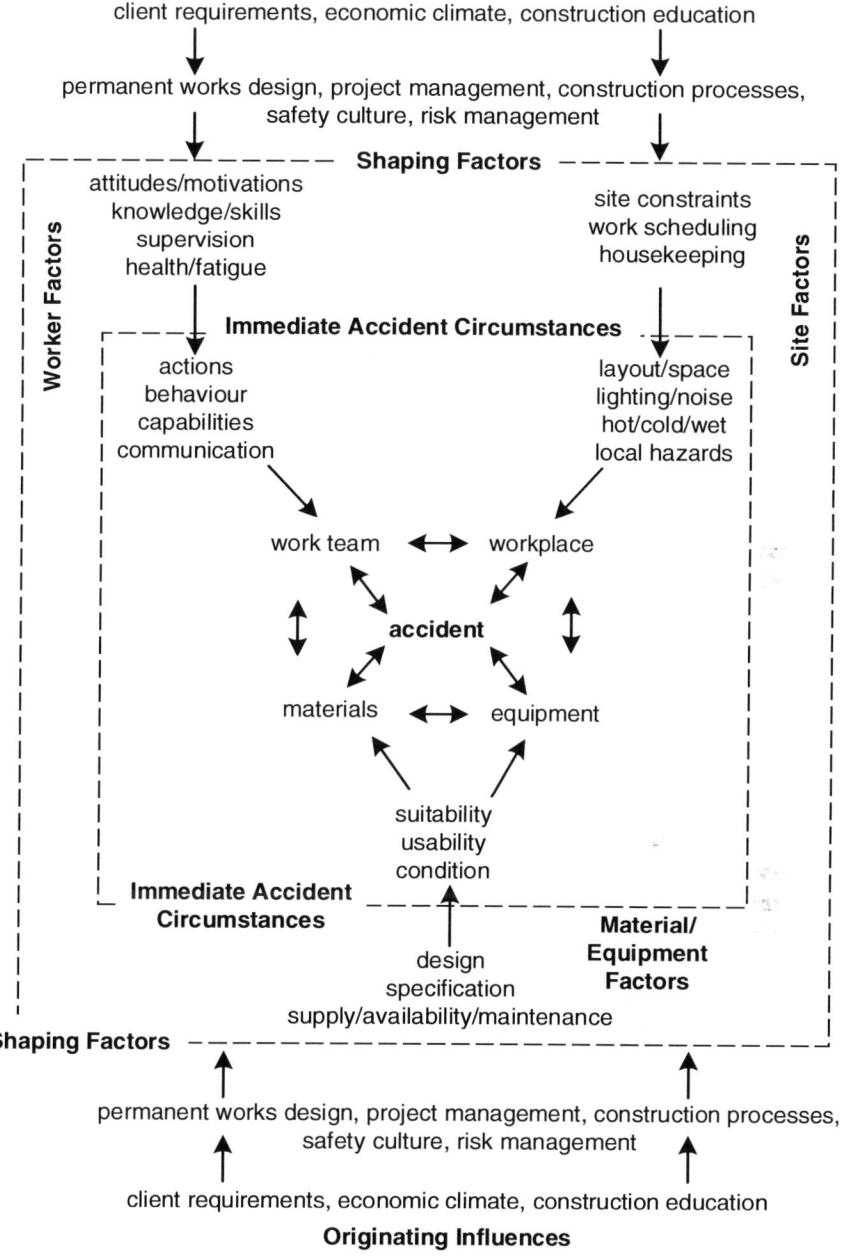

Originating Influences

client requirements, economic climate, construction education

permanent works design, project management, construction processes,
safety culture, risk management

Worker Factors

Shaping Factors

attitudes/motivations
knowledge/skills
supervision
health/fatigue

site constraints
work scheduling
housekeeping

Site Factors

Immediate Accident Circumstances

actions
behaviour
capabilities
communication

layout/space
lighting/noise
hot/cold/wet
local hazards

work team ⟷ workplace

accident

materials ⟷ equipment

suitability
usability
condition

**Immediate Accident
Circumstances**

**Material/
Equipment
Factors**

design
specification
supply/availability/maintenance

Shaping Factors

permanent works design, project management, construction processes,
safety culture, risk management

client requirements, economic climate, construction education

Originating Influences

Figure 6.1 Hierarchy of causal influences in construction accidents.
Source: Gibb *et al.* 2006.

circumstances'. These shortfalls occur due to the influence of the connected shaping factors. For instance, the actions, behaviours, capabilities and communications of the work team are shaped by their attitude, motivations, knowledge, skills, supervision, health and fatigue. Similarly, the level of hazards at the workplace is affected by site constraints, work scheduling and housekeeping. The suitability, usability, conditions and thereby safety of materials and equipment used on site are affected by their design, specification and supply/availability/maintenance. The shaping factors are affected by broader, distal factors, referred to as 'originating influences', such as the permanent works design, project management, construction processes, safety culture, risk management, client requirements, economic climate and construction education.

Figure 6.2 illustrates the extension of the Hierarchy of Causal Influences Model of Gibb *et al.* (2006) by incorporating primary and secondary psychological injuries and their causal factors, which were not part of the original model of Gibb *et al.* (2006). This produces a holistic incident causation model for construction, explaining the causes of physical injuries, illnesses, work-induced psychological disorders and injury-induced psychological disorders. The following sections explain the causes and their interactions that result in incidents.

Physical injuries and illnesses

Incidents that lead to fatalities, physical injuries or illnesses occur due to issues related to the: (1) workplace (layout/space, lighting/noise levels, hot/cold/wet conditions, local hazards); (2) materials and equipment (suitability, usability, condition); and/or (3) work team (actions, behaviour, capabilities, communications). These are referred to as 'immediate incident circumstances' for fatalities, injuries or illnesses.

These shortfalls occur due to the influence of the connected shaping factors. For instance, the level of hazards at the workplace is affected by site constraints, work scheduling and housekeeping. The suitability, usability, conditions and thereby safety of materials and equipment used on site are affected by their design, specification and supply/availability/maintenance. Similarly, safe/unsafe actions, safe/unsafe behaviour, safety capabilities and communications of the work team are shaped by their attitude towards safety, motivations to work safely, safety knowledge, skills, supervision, health and fatigue.

The shaping factors are influenced by broader, distal factors, referred to as 'originating influences', which are shown in two distinct boxes in the extreme left in the model. The first box concerns broader industry/client-level factors whilst the second box concerns design and construction organisational factors. The broader client/industry-level factors include the client requirements, economic climate and construction education. Client requirements for a project may dictate the level of hazards present in a project. For instance, an expectation to build the tallest building in the world will inevitably create the need for working at heights and thereby fall hazards. The economic climate that prevails at a

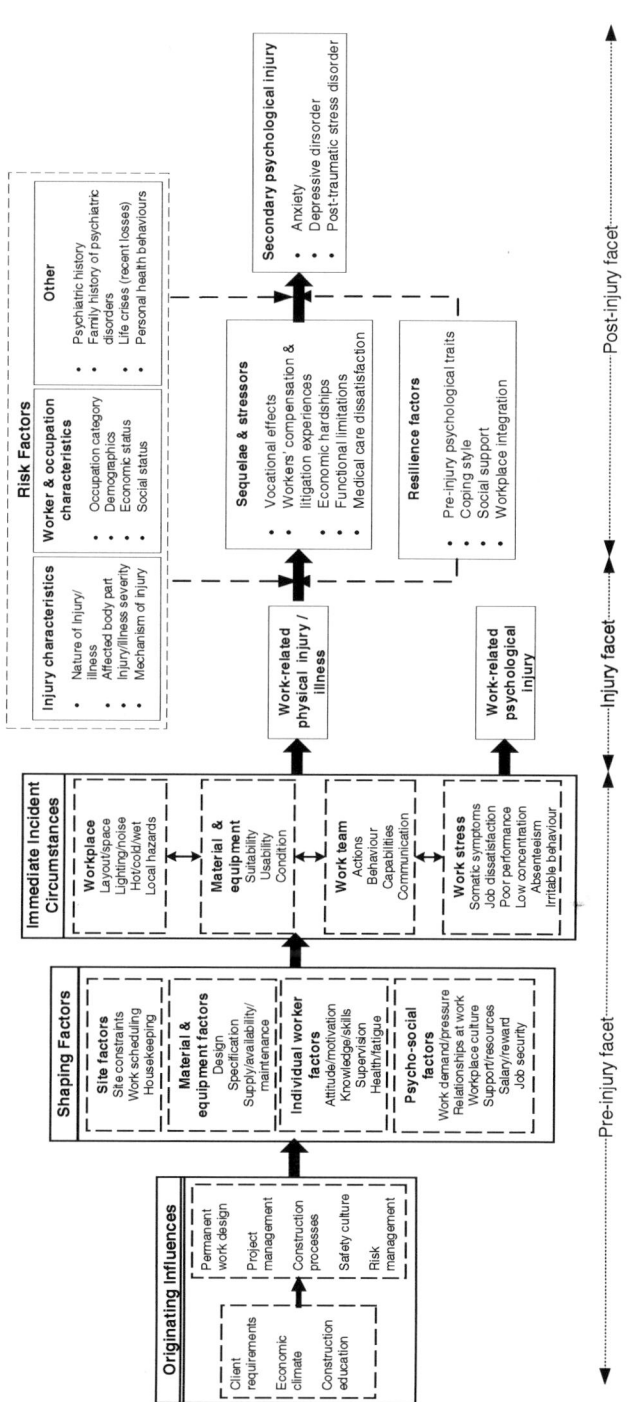

Figure 6.2 Construction incident causation model.

particular time drives the availability and the type of construction jobs in the market and the number of projects that an organisation has in hand. This may have an impact on safety implementations in projects. The quality of construction education in the country will influence the skills and competencies of professionals who work in the industry to manage safety. Design and construction organisational factors include design, project management, construction methods, safety culture and risk management. Permanent work design is a critical factor in creating or eliminating hazards in a project. Designs that consider safety during construction will create safe workplaces for workers and vice versa. Similarly, the construction and project management methods utilised by the builder influence the level of safety on site. Moreover, the prevalence of a strong safety culture and risk management regimes in a builder organisation will have a strong influence on safety climate on site. All these collectively influence the shaping factors.

Primary psychological injuries

The same originating influences discussed above form psycho-social conditions on sites that may be favourable or unfavourable to workers. Unfavourable psycho-social conditions lead to work stress, which is an immediate incident circumstance for work-induced psychological disorders (also referred to as primary psychological injuries).

The key psycho-social factors that cause work stress include: work demand/ pressure, relationships at work, workplace culture, support and resources, salary/ reward and job insecurity. Excessive work pressure, created by work overload and time pressure, causes enormous mental pressure for workers. Similarly, poor relationships with co-workers and supervisors and the prevalence of undesirable workplace culture such as discrimination, bullying and harassment add to the pressure. Lack of support from co-workers and supervisors and lack of resources to manage the work pressure will lead the accumulation of distress. Adding to this, not receiving adequate salary/wages and rewards for the work as well as job insecurity will worsen the mental distress. The simultaneous interactions of these poor psycho-social factors collectively result in excessive work stress.

Excessive work stress is manifested in multiple ways in workers, including: suffering somatic symptoms such as headaches, muscular pains, loss of appetite, insomnia and fatigue; job dissatisfaction; lack of concentration and poor decision making at work; lowered performance; absenteeism and irritable behaviour that damages relationships at work. Suffering excessive work stress for a prolonged period results in work-induced psychological disorders, namely depression, anxiety, suicidality and alcohol/substance abuse. Work stress also causes certain critical diseases such as hypertension, coronary heart disease, heart attacks and diseases of the digestive system.

Excessive work stress is a contributory factor for physical injuries too. Affected workers suffer from reduced motivation to work safely, lack of concentration on safety instructions/warning and poor performance. Moreover, pressure to finish work might make them take short-cuts, ignoring safety standards.

Secondary psychological injuries

Workers who suffer a work-related injury or illness may subsequently develop psychological disorders such as anxiety, depression, post-traumatic stress disorder or even suicidal thoughts due to the mental stress caused by negative changes to their physical, socio-economic and work status and the lack of support.

When a workplace injury or illness causes a permanent disability, impairment and/or chronic pain to a worker, it significantly disrupts and compromises work, social and personal life in the following ways, exerting enormous mental stress:

- It may force a change of job role, productivity loss, wage loss, and diminished job satisfaction and motivation and/or even loss of job or job security.
- It can lead to economic hardships and compromised quality of life for the entire family; losing the job or wages but at the same time the need for additional expenses for medical care and extra care may force workers to move/downgrade their primary residence and change their lifestyle to cut down on the cost of living, impacting on the quality of life of the entire family, including children's education and future.
- Workers' compensation (WC) insurance is meant to ease the economic hardship of permanently disabled/impaired workers by paying for all the medical expenses and compensating for lost wages and bodily damage. However, most WC policies cap the maximum amount of lost wages payable, leaving disabled or impaired workers in a net loss of earnings. Furthermore, workers may have to wait several months to receive wage replacement benefits and/or compensation from the insurer due to complex approval and processing. In both situations workers are left in economic hardship and distress.
- Delays of WC providers in paying compensation and medical expenses pose challenges for workers in obtaining appropriate healthcare. Patient satisfaction with healthcare provided through WC insurance appears to be generally lower than the general healthcare provided for non-occupational conditions.
- It causes lifelong limitations for performing daily activities such as household chores, fulfilling family responsibilities, and participating in recreational and social activities.

Some workers may be more vulnerable to SPIs than others because of the moderating effects of risk factors and resilient factors they are exposed to. Key risk factors include the psychiatric history of the worker and concurrently facing other life crises. Likewise, vital factors that help neutralise the stress are: the psychological trait of the worker, stress coping style, social support and workplace integration/support.

Practical implications

The findings discussed in the book, including the new incident causation model, have many practical implications for various stakeholders in the construction industry, namely: builders, incident investigators/WHS authorities, rehabilitation service providers, WHS policy makers/government authorities and WHS training providers/educational institutions.

The incident causation model, along with the theoretical insights discussed in the chapters of the book, provide a blueprint for builders on aspects that they should manage in their projects to eliminate workplace hazards – including psychological hazards – to minimise incidents such as fatalities, injuries, ill-health and mental illness to construction operatives. Moreover, most of the existing causation models focus on physical injuries and illnesses that affect operatives (blue-collar workers). The proposed model and the findings in Chapter 4 illuminate the psycho-social hazards that affect professionals (white-collar workers), enabling construction organisations to manage the well-being of their professional workforce. This will eventually assist in improving the productivity of their organisations and the industry in general.

Incident investigators traditionally evaluate the physical workplace and behaviour of work teams in establishing evidence/causes for incidents. The proposed model adds two more dimensions to their investigation template. First, it shows the influence of work stress on incidents and the indicators that may be used to assess whether work stress has been a contributory factor in an incident. Then, along with the theoretical model discussed in Chapter 4, it provides a list of indicators for a template to evaluate the level of work stress in construction organisations and sites to suggest de-stressing needs for improved productivity. In summary, it assists in conducting comprehensive investigations that cover both physical and psycho-social causes of incidents.

The book has, for the first time, introduced SPIs to the construction industry/ literature along with their causal and risk factors, as shown in the new incident causation model. This new evidence can inform rethinking of rehabilitation programmes. Injured workers are signed up for rehabilitation programmes and a rehabilitation consultant works with the worker, medical doctor, the employer and the insurer to ensure the worker returns to work as quick as possible. The consultant conventionally focuses on physical rehabilitation. The model and the information about the causal and risk factors of SPI can guide rehabilitation consultants with a list of matters related to sequelae of work injuries and ill-nesses, which they should deal with for ensuring the psychological well-being of the injured workers in addition to physical rehabilitation. This will make the rehabilitation programmes more comprehensive and more beneficial for affected workers.

The current workers' compensation insurance scheme in Australia, and in many other countries, does not cover SPIs. Construction workers' unions have been lobbying for the inclusion of SPIs in the scheme but without success so far because there is no scientific evidence to establish that injured construction workers are vulnerable to subsequent psychological disorders. The new causation model and the information discussed in Chapter 5 of this book can provide scientific evidence in this regard to inform policymakers and government authorities/legislators of the importance of incorporating secondary psychological injuries in the workers' compensation system as well as tightening regulations concerning fair compensation, adequate medical support and workplace support in a timely manner for injured workers.

Most of the textbooks used by tertiary institutions that provide construction management education and/or WHS training largely cover the causes of incidents that result in physical injuries, illnesses and fatalities. With the new causation model and the new insights discussed in the chapters of this book, the focus can be broadened to include: (1) the influence of psycho-social factors on incidents that result in fatalities, injuries and illnesses; (2) work-induced stress and its impact on well-being; and (3) post-injury psychological disorders and their work relatedness. Such a broadening effort would eventually trigger change in the mindset and thinking about WHS at a higher level (at the originating level) in the long run.

The potential changes to the way WHS is managed by different stakeholders, as discussed above, would collectively contribute to reducing incidents in the construction industry. The current incident rate in the construction industry creates a distressing socio-economic burden globally. The positive changes would help reduce this burden and improve the social sustainability worldwide.

Methodological insights

The book demonstrates the application of new methods for construction safety research. In a broader perspective, it showcased the application of data mining and analytics in construction safety research, which was an underexplored area. Specifically, it demonstrates the use of three data mining and analytics techniques, namely multiple correspondence analysis, CHAID tree analysis and neural network modelling, resulting in the lessons described below.

- Multiple correspondence analysis is a powerful tool for analysing datasets with multiple categorical variables of qualitative data that are described with multiple measurement scales/continuum. It also provides graphical representation of outputs. However, this technique is hardly used in construction research, possibly due to the lack of previous work that demonstrates its application. This book applied the technique on two different scenarios to showcase its methodical application and the interpretation of results, which should provide insights to other researchers.
- Classification trees is a versatile technique that can handle skewed datasets, and as such is widely used in many fields such as medicine, computer science and psychology for causality as well as association analysis. It uses different machine learning algorithms for creating classification models, depending on the type of data used, whether categorical or scale/metric. The chi-squared

automatic interaction detection (CHAID) algorithm is suitable for creating classification models for categorical explanatory and outcome variables. As far as the author is aware, this tool has been rarely used in construction research despite its versatility. This book demonstrated its application twice and thereby should provide insights for other researchers

- Despite the increased use of neural networks in many fields, to date it is largely an overlooked method in construction health, safety and well-being research, possibly due to the lack of previous research that showcases the application effectiveness in this domain. This book provides fresh insights and an exemplary application for construction health and safety researchers.

Future research

The main theme of this book has been the application of data mining and analytics for improving health, safety and well-being in the construction industry. It was realised while writing the book that there is ample potential for future research, both by continuing the same theme and by starting on new related topics. Hence, the following topics and themes are suggested for future research.

- The various research topics discussed in this book leveraged the workers' compensation data collected by Safe Work Australia. This is just one type of data about construction incidents collected by WHS authorities such as Safe Work Australia, and there are more types available about incidents; to mention two examples, incident investigation reports and coroner reports. Future research aimed at the application of text analytics on these sets of data would reveal further new knowledge to improve health, safety and well-being in construction.
- Many other industries use wearable technologies to constantly monitor and improve safety, health, well-being and ergonomics of workers. However, the construction industry is lagging in this aspect despite it having one of the worst incident rates, warranting an increased application rate of these devices. Further research is suggested to demonstrate the utilisation and benefits of apt wearable technologies for construction workers for different hazard scenarios. Some examples of scenarios in which wearable technologies can be investigated are as follows:
    - Mental stress and fatigue are some of the main causes of serious accidents and fatalities in construction. Utilisation of wearable technologies to monitor stress and fatigue levels and their impact on safety and well-being can be investigated on construction sites. The study can be further extended to compare the return on investment on these devices and programmes.
    - The application of wearable technologies to improve ergonomics of construction workers can be investigated through different tasks that involve repetitive movements, manual lifting, body twisting and

bending, and awkward postures. The data collected may guide the development of ergonomically safe work practices for different occupation trades.

- Research has demonstrated that poor mental health is quite rampant among construction professionals, which affects the productivity and profitability of the overall industry. Identifying and supporting professionals with elevated stress, anxiety and depression, who otherwise will not disclose their conditions due to stigma, is essential for a healthy industry. The practicality and methodology of using emotional well-being wearables for improving the psychological health of construction professionals may be investigated both on site and at office levels.

- Often, work-related illnesses are believed to be caused by exposure to physical hazards in construction. Accordingly, prevention measures are suggested which are also aimed at improving the physical workplace. However, serious illnesses such as cardiovascular diseases and the diseases of the digestive system can be caused by work stress, particularly among white-collar professionals in the construction industry. There is a dearth of studies that investigate the associations between work stress and the prevalence of serious illnesses among construction professionals. Future research may be undertaken on this topic.

- Secondary psychological injuries (SPIs) are a largely underexplored topic in construction. This book introduced this terminology in the construction literature for the first time. Further research is needed to explore this aspect of WHS in multiple perspectives. Some suggested directions/perspectives are as follows:

  - The experiences and satisfaction of injured workers with workers' compensation, medical care and rehabilitation programmes may be explored to produce scientific evidence to inform policy and legislative changes.
  - Development of a mobile app and a dashboard for monitoring psychological health of injured workers. The technology can capture the experiences/satisfaction of injured workers with various systems such as workers' compensation, medical care and personal/social life and how it impacts on their mental health. This can be used by rehabilitation practitioners and/or medical practices to proactively prevent SPIs.

References

Abdelhamid, T.S. and Everett, J.G. (2000). Identifying root causes of construction accidents. *Journal of Construction Engineering and Management*, 126(1): 52–60.

Carter, G. and Smith, S.D. (2006). Safety hazard identification on construction projects. *Journal of Construction Engineering and Management*, 132(2): 197–205.

Gibb, A., Haslam, R., Gyi, D., Hide, S., and Duff, R. (2006). What causes accidents? *Proceedings of the ICE – Civil Engineering*, 159(6): 46–50.

Heinrich, H.W. (1931). *Industrial Accident Prevention: A Scientific Approach*. New York: McGraw-Hill.

Larsson, S., Pousette, A., and Törner, M. (2008). Psychological climate and safety in the construction industry-mediated influence on safety behaviour. *Safety Science*, 46: 405–412.

Leung, M.Y., Chan, Y.S., and Yuen, K.W. (2010). Impacts of stressors and stress on the injury incidents of construction workers in Hong Kong. *Journal of Construction Engineering and Management*, 136: 1093–1103.

Petersen, D. (1971). *Techniques of Safety Management*. New York: McGraw-Hill.

Petersen, D. (1982). *Human Error – Reduction and Safety Management*. New York: STPM Press.

Reason, J. (1990). *Human Error*. Cambridge: Cambridge University Press.

Siu, O., Phillips, D., and Leung, T. (2004). Safety climate and safety performance among construction workers in Hong Kong: The role of psychological strains as mediators. *Accident Analysis & Prevention*, 36(3): 359–366.

Index

Page numbers in **bold** denote tables, those in *italics* denote figures.